FOURTH EDITION

A CONCISE
INTRODUCTION TO
PURE
MATHEMATICS

FOURTH EDITION

A CONCISE INTRODUCTION TO PURE MATHEMATICS

MARTIN LIEBECK

CRC Press
Taylor & Francis Group
Boca Raton London New York

CRC Press is an imprint of the
Taylor & Francis Group, an **informa** business

A CHAPMAN & HALL BOOK

CRC Press
Taylor & Francis Group
6000 Broken Sound Parkway NW, Suite 300
Boca Raton, FL 33487-2742

© 2016 by Taylor & Francis Group, LLC
CRC Press is an imprint of Taylor & Francis Group, an Informa business

No claim to original U.S. Government works

ISBN 13: 978-1-4987-2292-6 (pbk)

Library of Congress Cataloging-in-Publication Data

Liebeck, M. W. (Martin W.), 1954-
   A concise introduction to pure mathematics / Martin Liebeck. -- Fourth edition.
      pages cm
   "A CRC title."
   Includes bibliographical references and index.
   ISBN 978-1-4987-2292-6 (alk. paper)
    1. Logic, Symbolic and mathematical. 2. Mathematics. I. Title.

QA9.L478 2016
510--dc23
                                                    2015025649

Visit the Taylor & Francis Web site at
http://www.taylorandfrancis.com

and the CRC Press Web site at
http://www.crcpress.com

To Ann, Jonny and Matthew

Contents

Foreword

One of the great difficulties in teaching undergraduate mathematics at universities in the United States is the great gap between teaching students a set of algorithms (which is very often the bulk of what is learned in calculus) and convincing students of the power, beauty and fun of the basic concepts in mathematics.

Martin Liebeck's book, *A Concise Introduction to Pure Mathematics, Fourth Edition*, is one of the best I have seen at filling this gap. In addition to preparing students to go on into mathematics, it is also a wonderful choice for a student who will not necessarily go on in mathematics but wants a gentle but fascinating introduction into the culture of mathematics. Liebeck starts with the basics and introduces number systems. In particular he discusses the real numbers and complex numbers. He shows how these concepts are natural and important in solving natural problems. Various topics in analysis, geometry, number theory and combinatorics are introduced and are shown to be fun and beautiful. Starting from scratch, Liebeck develops interesting results which hopefully will intrigue the student and give encouragement to continue to study mathematics.

This book will give a student the understanding to go on to further courses in abstract algebra and analysis. The notion of a proof will no longer be foreign, but also mathematics will not be viewed as some abstract black box. At the very least, the student will have an appreciation of mathematics.

As usual, Liebeck's writing style is clear and easy to read. This is a book that could be read by a student on his or her own. There is a wide selection of problems ranging from routine to quite challenging.

While there is a difference in mathematical education between the U.K. and the U.S., this book will serve both groups of students extremely well.

Professor Robert Guralnick
Chair of Mathematics Department
University of Southern California
Los Angeles, California

Preface

I can well remember my first lecture as a mathematics undergraduate, back in the olden days. In it, we were told about something called "Russell's Paradox" — does the set consisting of all sets which do not belong to themselves belong to itself? — after which the lecturer gave us some rules called the "Axioms of Set Theory." I came out of the lecture somewhat baffled. The second lecture, in which we were informed that "a_n tends to l if, for every $\varepsilon > 0$, there exists N such that for all $n \geq N$, $|a_n - l| < \varepsilon$," was also a touch bewildering. In fact, the lecturers were pretty good, and bafflement and bewilderment eventually gave way to understanding, but nevertheless it was a fairly fierce introduction to the world of university pure mathematics.

Nowadays we university lecturers are less fierce, and mathematics courses tend to start with a much gentler introduction to pure mathematics. I gave such a course at Imperial College for several years to students in the first term of the first year of their degree (generally in mathematics, or some joint degree including mathematics). This book grew out of that course. As well as being designed for use in a first university course, the book is also suitable for self-study. It could, for example, be read by students between school and university, or indeed by anybody with a reasonable background in school mathematics.

One of my aims is to provide a robust bridge between school and university mathematics. For a number of the topics covered, students may well have studied some of the basic material on this topic at school, but this book will generally take the topic much further, in a way that is interesting and stimulating (at least to me). For example, many will have come across the method of mathematical induction, and used it to solve some simple problems, like finding a formula for the sum $1 + 2 + 3 + \ldots + n$. But I doubt that many have seen how induction can be used to study solid objects whose faces all have straight edges, and to show that the only so-called regular solids are the famous five "Platonic solids" (the cube, tetrahedron, octahedron, icosahedron and dodecahedron), as is done in Chapter 9.

I generally enjoy things more if they come in bite-sized pieces, and accordingly I have divided the book into 26 short chapters. Each chapter ends with a selection of exercises, ranging from routine calculations to some quite challenging problems.

When starting to study pure mathematics at university, students often have a refreshing sense of "beginning all over again." Basic structures, like the real numbers, the integers, the rational numbers and the complex numbers, must be defined and studied from scratch, and even simple and obvious-looking statements about them must be proved properly. For example, it probably seems obvious that if n is an integer (i.e., one of the whole numbers $0, 1, -1, 2, -2,$ $3, -3$ and so on), and n^2 is odd, then n must also be odd. But how can we write down a rigorous proof of this fact? Methods for writing down proofs of this and many other simple facts form one of the themes of Chapter 1, along with a basic introduction to the language of sets.

In Chapter 2, I define and begin to study three of the basic number systems referred to in the previous paragraph: the real numbers (which we start off by thinking of as points on an infinite straight line — the "real line"); the integers; and the rational numbers (which are the fractions $\frac{m}{n}$, where m and n are integers). It takes some effort to prove that there is at least one real number that is not rational — a so-called irrational number — but once this is done, one can see quite easily that there are many irrational numbers. Indeed, by Chapter 21 we shall understand the strange fact that, in a very precise sense, there are "more" irrational numbers than rational numbers (even though there are infinitely many of each).

In studying properties of the system of real numbers, it is sometimes helpful to have ways of thinking of them that are different from just "points on the real line." In Chapter 3, I introduce the familiar decimal notation for real numbers, which provides a visual way of writing them down and can be useful in their general study. Chapters 4 and 5 carry on with our basic study of the real numbers.

In Chapter 6, I bring our last important number system into the action — the complex numbers. Students may well have met these before. We begin by introducing a symbol i, and defining $i^2 = -1$. A general complex number is a symbol of the form $a + bi$, where a and b are real numbers. We soon find that using complex numbers we can write down solutions of all quadratic equations, and then proceed to study other equations like $x^n = 1$. We also find some beautiful links between complex numbers and geometry in the plane. Chapter 7 takes the theory of equations much further. Solving quadratics is probably very familiar, but much less well known is the method for solving cubic equations given in this chapter. We then look at general polynomial equations (i.e., equations of the form $x^n + a_{n-1}x^{n-1} + \ldots + a_1 x + a_0 = 0$) and explore the amazing fact that every such equation has solutions which are complex numbers (known as the Fundamental Theorem of Algebra).

I have already mentioned the method of proof by mathematical induction, which is introduced in Chapter 8. This is a technique for proving statements involving a general positive integer n, such as "the sum of the first n odd positive integers is equal to n^2," or "the number of regions formed by n straight lines drawn in the plane, no two parallel and no three going through the same point, is equal to $\frac{1}{2}(n^2 + n + 2)$." The method of induction is actually rather more powerful than first meets the eye, and Chapter 9 is devoted to the proof by induction of an elegant result, known as Euler's formula, about the relationship between the numbers of corners, edges and faces of a solid object, whose faces all have straight edges. Euler's formula has all sorts of uses. For example, if you want to make a football by sewing together hexagonal and pentagonal pieces of leather, in such a way that each corner lies on three edges, then the formula implies that you will need exactly 12 pentagonal pieces, no more and no less. I could not resist going further in this chapter and showing how to use Euler's formula to study the famous Platonic solids mentioned earlier.

Chapters 10 through 14 are all about possibly the most fascinating number system of all: the integers. Students will know what a prime number is — an integer greater than 1 that is only divisible by 1 and itself — and are quite likely aware of the fact that every integer greater than 1 is equal to a product of prime numbers, although this fact requires a careful proof. Much more subtle is the fact that such a prime factorization is unique — in other words, given an integer greater than 1, we can express it as a product of prime numbers in only one way. "Big deal! So what?" I hear you say. Well, yes, it is a big deal (so big that this result has acquired the grandiose title of "The Fundamental Theorem of Arithmetic"), and after proving it I try to show its significance by using it in the study of a number of problems; for instance, apart from 1 and 0, are there any squares that differ from a cube by just 1?

Chapter 15 contains an amazing application of some of the theory of prime numbers developed in the previous chapters. This application concerns some very clever secret codes that are used every day for the secure electronic transmission of sensitive information — one of today's most spectacular "real-world" applications of pure mathematics.

Chapter 16 is about methods of counting things. For example, suppose I have given the same lecture course for the last 16 years, and tell 3 jokes each year. I never tell the same set of 3 jokes twice. At least how many jokes do I know? To solve this and other important counting problems, we introduce binomial coefficients, which leads us into the binomial and multinomial theorems.

After a little formal theory of sets and relations in Chapters 17 and 18, I introduce functions in Chapter 19, and develop some of the delights of the theory of an especially interesting and important class of functions called "permutations" in Chapter 20. Then comes Chapter 21, in which I address some fascinating questions about infinite sets. When can we say that two infinite sets have the same "size"? Can we ever say that one infinite set has bigger "size" than another? These questions are answered in a precise and rigorous way, and some of the answers may appear strange at first sight; for example, the set of all integers and the set of all rational numbers have the same size, but the set of all real numbers has greater size than these. Chapter 21 closes with a beautifully subtle result that tells us that an infinite set always has smaller size than the set of all its subsets. The proof of this is based on the argument of Russell's Paradox — which brings me back to where I started ....

The next three chapters have a somewhat different flavour to the rest of the book. In them I introduce a topic known as mathematical analysis, which is the study of the real numbers and functions defined on them. Of course I can't cover very much of the subject — that would require several more books — but I do enough to prove several interesting results and to fill in one or two gaps in the preceding chapters. The point is that with our somewhat naive understanding of the real numbers up to here, it is difficult to see how to prove even such basic properties as the fact that every positive real number has a square root, a cube root and so on. The material in Chapters 22–24 is sufficient at least to prove this fact, and also to do some other interesting things, such as proving a special case of the famous Fundamental Theorem of Algebra.

In the final two chapters of the book I introduce another very different kind of mathematics — the theory of groups, which is part of a huge area known as abstract algebra. Groups are defined as sets of objects (they could be numbers, or functions, or anything really), together with a rule for combining any two objects to get another one, and this rule is subject to four clearly defined assumptions, called the "axioms" of group theory. The game is to see what one can deduce just using the axioms. Fortunately the subject is more than just a game, and there are many beautiful examples and applications.

Let me now offer some comments on designing a course based on the book. Crudely speaking, the book can be divided into six fairly independent sections, with the following "core" chapters:

Introduction to number systems: Chapters 1, 2, 3, 4, 5, 6, 8

Theory of the integers: Chapters 10, 11, 13, 14

Introduction to discrete mathematics: Chapters 16, 17, 19, 20

Functions, relations and countability: Chapters 18, 19, 21

Introduction to analysis: Chapters 22, 23, 24

Introduction to abstract algebra: Chapters 25, 26

One could design a one- or two-semester course in a number of ways. For example, if the emphasis is to be on discrete mathematics, the core chapters to use from the first section would be 1, 2 and 8, and all the other sections except the last two would be core; the last section on abstract algebra would also be a natural addition to such a course. On the other hand, if the course is intended to prepare students more for a future course in analysis, one should use all the chapters in the first, fourth and fifth sections. Overall, I would recommend incorporating at least the first four sections into your course — it works well!

I would like to express my thanks to my late father, Dr. Hans Liebeck, who read the entire manuscripts of the first two editions and suggested many improvements, as well as saving me from a number of embarrassing errors. Sadly, I can no longer claim that any errors that remain are his responsibility. And, finally, I thank generations of students at Imperial who have sat through my lectures and have helped me to hone the course into the sleek monster that has grown into this book.

New Features of the Fourth Edition

This fourth edition contains several substantial additions to the third edition. First, I have included two new chapters at the end to serve as an introduction to the topic of abstract algebra. This is a big subject which is often introduced in a course of its own at the undergraduate level, but I believe it fits quite well into the framework of this book. For one thing, it is a topic that one can begin to read about from scratch, without needing to know too much other stuff; on the other hand, many of the examples and applications are closely connected with other parts of the book, particularly the chapters on number systems, prime numbers, congruence and permutations. It also gives an introduction to "abstract" reasoning in mathematics, where one is allowed only to use a set of axioms and nothing else, and often students find this a new and exciting challenge.

I have also added new material in a number of other chapters: on inequalities in Chapters 5 and 8; on counting methods in Chapter 16; and on the Inclusion–Exclusion Principle and Euler's ϕ-function in Chapter 17. There are also lots of new exercises, and, as in the previous edition, I have included solutions to the odd-numbered ones.

Chapter 1

Sets and Proofs

This chapter contains some introductory notions concerning the language of sets, and methods for writing proofs of mathematical statements.

Sets

We shall think of a *set* as simply a collection of objects, which are called the *elements* or *members* of the set. There are a number of ways of describing a set. Sometimes the most convenient way is to make a list of all the objects in the set and put curly brackets around the list. Thus, for example,

$\{1,3,5\}$ is the set consisting of the objects 1, 3 and 5.

$\{\text{Fred}, \text{dog}, 1.47\}$ is the set consisting of the objects Fred, dog and 1.47.

$\{1, \{2\}\}$ is the set consisting of two objects, one being the number 1 and the other being the set $\{2\}$.

Often, however, this is not a convenient way to describe our set. For example, the set consisting of all the people who live in Denmark is for most purposes best described by precisely this phrase (i.e., "the set of all people who live in Denmark"); it is unlikely to be useful to describe this set in list form $\{\text{Sven}, \text{Inge}, \text{Jesper}, \ldots\}$. As another example, the set of all real numbers whose square is less than 2 is neatly described by the notation

$$\{x \mid x \text{ a real number}, x^2 < 2\}.$$

(This is to be read: "the set of all x such that x is a real number and $x^2 < 2$." The symbol "$\mid$" is the "such that" part of the phrase.) Likewise,

$$\{x \mid x \text{ a real number}, x^2 - 2x + 1 = 0\}$$

denotes the set consisting of all real numbers x such that $x^2 - 2x + 1 = 0$.

As a convention, we define the *empty set* to be the set consisting of no objects at all, and denote the empty set by the symbol $\emptyset$.

If S is a set, and s is an element of S (i.e., an object that belongs to S), we write

$$s \in S$$

and say s *belongs to* S. If some other object t does not belong to S, we write

$$t \notin S.$$

For example,

$1 \in \{1, 3, 5\}$ but $2 \notin \{1, 3, 5\}$,

if $S = \{x \mid x$ a real number, $0 \le x \le 1\}$, then $1 \in S$ but Fred $\notin S$,

$\{2\} \in \{1, \{2\}\}$ but $2 \notin \{1, \{2\}\}$,

$1 \notin \emptyset$.

Two sets are defined to be equal when they consist of exactly the same elements; for example,

$\{1, 3, 5\} = \{3, 5, 1\} = \{1, 5, 1, 3\}$,

$\{x \mid x$ a real number, $x^2 - 2x + 1 = 0\} = \{1\}$,

$\{x \mid x$ a real number, $x^2 + 1 = 0\} =$ the set of female popes $= \emptyset$.

We say a set T is a *subset* of a set S if every element of T also belongs to S (i.e., T consists of some of the elements of S). We write $T \subseteq S$ if T is a subset of S, and $T \nsubseteq S$ if not. For example, if $S = \{1, \{2\}, \text{cat}\}$, then

$$\{\text{cat}\} \subseteq S, \ \ \{\{2\}\} \subseteq S, \ \ \{2\} \nsubseteq S.$$

As another example, the subsets of $\{1, 2\}$ are

$$\{1, 2\}, \{1\}, \{2\}, \emptyset.$$

(By convention, $\emptyset$ is a subset of every set.)

This is all we shall need about sets for the time being. This topic will be covered somewhat more formally in Chapter 17.

Proofs

Consider the following mathematical statements:

(1) The square of an odd integer is odd. (By an *integer* we mean a whole number, i.e., one of the numbers $\ldots, -2, -1, 0, 1, 2, \ldots$.)

(2) No real number has square equal to -1.

(3) Every positive integer is equal to the sum of two integer squares. (The integer squares are $0, 1, 4, 9, 16, 25$, and so on.)

Each of these statements is either true or false. Probably you have quickly formed an opinion on the truth or falsity of each, and regard this as "obvious" in some sense. Nevertheless, to be totally convincing, you must provide clear, logical proofs to justify your opinions.

To clarify what constitutes a proof, we need to introduce a little notation. If P and Q are statements, we write

$$P \Rightarrow Q$$

to mean that statement P implies statement Q. For example,

$x = 2 \Rightarrow x^2 < 6,$

it is raining $\Rightarrow$ the sky is cloudy.

Other ways of saying $P \Rightarrow Q$ are:

if P then Q (e.g., if $x = 2$ then $x^2 < 6$);

Q if P (e.g., the sky is cloudy if it is raining);

P only if Q (e.g., $x = 2$ only if $x^2 < 6$; it rains only if the sky is cloudy).

Notice that $P \Rightarrow Q$ does *not* mean that also $Q \Rightarrow P$; for example, $x^2 < 6 \not\Rightarrow x = 2$ (where $\not\Rightarrow$ means "does not imply"). However, for some statements P, Q, it *is* the case that both $P \Rightarrow Q$ and $Q \Rightarrow P$; in such a case we write $P \Leftrightarrow Q$, and say "P if and only if Q." For example,

$x = 2 \Leftrightarrow x^3 = 8,$

you are my wife if and only if I am your husband.

The *negation* of a statement P is the opposite statement, "not P," written as the symbol $\bar{P}$. Notice that if $P \Rightarrow Q$ then also $\bar{Q} \Rightarrow \bar{P}$ (since if $\bar{Q}$ is true then P cannot be true, as $P \Rightarrow Q$).

For example, if P is the statement $x = 2$ and Q the statement $x^2 < 6$, then $P \Rightarrow Q$ says "$x = 2 \Rightarrow x^2 < 6$," while $\bar{Q} \Rightarrow \bar{P}$ says "$x^2 \geq 6 \Rightarrow x \neq 2$." Likewise, for the other example above we have "the sky is not cloudy $\Rightarrow$ it is not raining."

Perhaps labouring the obvious, let us now make a list of the deductions that can be made from the implication "it is raining $\Rightarrow$ the sky is cloudy," given various assumptions:

Assumption	Deduction
it is raining	sky is cloudy
it is not raining	no deduction possible
sky is cloudy	no deduction possible
sky is not cloudy	it is not raining

Now let us put together some examples of proofs. In general, a proof will consist of a series of implications, proceeding from given assumptions, until the desired conclusion is reached. As we shall see, the logic behind a proof can take several different forms.

Example 1.1

Suppose we are given the following three facts:
   (a) I will be admitted to Greatmath University only if I am clever.
   (b) If I am clever then I do not have to work hard.
   (c) I have to work hard.
What can be deduced?

Answer Write G for the statement "I will be admitted to Greatmath University," C for the statement "I am clever," and W for the statement "I have to work hard." Then (a) says $G \Rightarrow C$, and (b) says $C \Rightarrow \bar{W}$. Hence,

$$W \Rightarrow \bar{C} \quad \text{and} \quad \bar{C} \Rightarrow \bar{G}.$$

Since W is true by (c), we deduce that $\bar{G}$ is true, i.e., I will not be admitted to Greatmath University (thank goodness).

Example 1.2

In this example we prove statement (1) from the previous page: the square of an odd integer is odd.

PROOF Let n be an odd integer. Then n is 1 more than an even integer, so $n = 1 + 2m$ for some integer m. Therefore, $n^2 = (1 + 2m)^2 = 1 + 4m + 4m^2 = 1 + 4(m + m^2)$. This is 1 more than $4(m + m^2)$, an even number, hence n^2 is odd. ∎

Formally, we could have written this proof as the following series of implications:

$$n \text{ odd} \Rightarrow n = 1 + 2m \Rightarrow n^2 = 1 + 4(m + m^2) \Rightarrow n^2 \text{ odd}.$$

However, this is evidently somewhat terse, and such an approach with more complicated proofs quickly leads to unreadable mathematics; so, as in the above proof, we insert words of English to make the proof readable, including words like "hence," "therefore," "then" and so on, to take the place of implication symbols.

Note The above proof shows rather more than just the oddness of n^2: it shows that the square of an odd number is always 1 more than a multiple of 4, i.e., is of the form $1 + 4k$ for some integer k.

The proofs given for Examples 1.1 and 1.2 could be described as *direct* proofs in that they proceed from the given assumptions directly to the conclusion via a series of implications. We now discuss two other types of proof, both very commonly used.

The first is *proof by contradiction*. Suppose we wish to prove the truth of a statement P. A proof by contradiction would proceed by first assuming that P is false — in other words, assuming $\bar{P}$. We would try to deduce from this a statement Q that is palpably false (for example, Q could be the statement "$0 = 1$" or "Liebeck is the pope"). Having done this, we have shown

$$\bar{P} \Rightarrow Q.$$

Hence also $\bar{Q} \Rightarrow P$. Since we know Q is false, $\bar{Q}$ is true, and hence so is P, so we have proved P, as desired.

The next three examples illustrate the method of proof by contradiction.

Example 1.3
Let n be an integer such that n^2 is a multiple of 3. Then n is also a multiple of 3.

PROOF   Suppose n is not a multiple of 3. Then when we divide n by 3, we get a remainder of either 1 or 2; in other words, n is either 1 or 2 more than a multiple of 3. If the remainder is 1, then $n = 1 + 3k$ for some integer k, so

$$n^2 = (1 + 3k)^2 = 1 + 6k + 9k^2 = 1 + 3\left(2k + 3k^2\right).$$

But this means that n^2 is 1 more than a multiple of 3, which is false, as we are given that n^2 is a multiple of 3. And if the remainder is 2, then $n = 2 + 3k$ for some integer k, so

$$n^2 = (2 + 3k)^2 = 4 + 12k + 9k^2 = 1 + 3\left(1 + 4k + 3k^2\right),$$

which is again false as n^2 is a multiple of 3.

Thus we have shown that assuming n is not a multiple of 3 leads to a false statement. Hence, as explained above, we have proved that n is a multiple of 3.   ∎

Usually in a proof by contradiction, when we arrive at our false statement Q, we simply write something like "this is a contradiction" and stop. We do this in the next proof.

Example 1.4
No real number has square equal to -1.

PROOF   Suppose the statement is false. This means that there *is* a real number, say x, such that $x^2 = -1$. However, it is a general fact about real numbers that the square of any real number is greater than or equal to 0 (see Chapter 5, Example 5.2). Hence $x^2 \geq 0$, which implies that $-1 \geq 0$. This is a contradiction.   ∎

Example 1.5
Prove that $\sqrt{2} + \sqrt{6} < \sqrt{15}$.

PROOF   Let me start by giving a non-proof:

$$\sqrt{2} + \sqrt{6} < \sqrt{15} \Rightarrow \left(\sqrt{2} + \sqrt{6}\right)^2 < 15$$
$$\Rightarrow 8 + 2\sqrt{12} < 15 \Rightarrow 2\sqrt{12} < 7 \Rightarrow 48 < 49.$$

The last statement $(48 < 49)$ is true, so why is this not a proof? Because the implication is going the wrong way — we have shown that if P is the statement we want to prove, and Q is the statement that $48 < 49$, then $P \Rightarrow Q$; but this tells us nothing about the truth or otherwise of P.

A cunning change to the above false proof gives a correct proof, by contradiction. So assume the result is false; i.e., assume that $\sqrt{2} + \sqrt{6} \geq \sqrt{15}$. Then

$$\sqrt{2} + \sqrt{6} \geq \sqrt{15} \Rightarrow \left(\sqrt{2} + \sqrt{6}\right)^2 \geq 15$$
$$\Rightarrow 8 + 2\sqrt{12} \geq 15 \Rightarrow 2\sqrt{12} \geq 7 \Rightarrow 48 \geq 49,$$

which is a contradiction. Hence we have proved that $\sqrt{2} + \sqrt{6} < \sqrt{15}$.
∎

The other method of proof we shall discuss is actually a way of proving statements are false — in other words, *disproving* them. We call the method *disproof by counterexample*. It is best explained by examples.

Example 1.6
Consider the following two statements:

(a) All men are Chinese.

(b) Every positive integer is equal to the sum of two integer squares.

As the reader will have cleverly spotted, both these statements are false. To disprove (a), we need to prove the negation, which is "not all men are Chinese," or equivalently, "there exists a man who is not Chinese"; this is readily done by simply displaying one man who is not Chinese — this man will then be a *counterexample* to statement (a). The point is that to disprove (a), we do not need to consider *all* men, we just need to produce a single counterexample.

Likewise, to disprove (b) we just need to provide a single counter-example — that is, a positive integer that is *not* equal to the sum of two squares. The number 3 fits the bill nicely.

Quantifiers

I will conclude the chapter by slightly formalising some of the discussion we have already had about proofs.

Consider the following statements:

(1) There is an integer n such that $n^3 = -27$.

(2) For some integer x, $x^2 = -1$.

(3) There exists a positive integer that is not equal to the sum of three integer squares.

Each of these statements has the form: "there exists some integer with a certain property." This type of statement is so common in mathematics that we represent the phrase "there exists" by a special symbol, namely $\exists$. So, writing $\mathbb{Z}$ for the set of all integers, the above statements can be rewritten as follows:

(1) $\exists n \in \mathbb{Z}$ such that $n^3 = -27$.

(2) $\exists x \in \mathbb{Z}$ such that $x^2 = -1$.

(3) $\exists x \in \mathbb{Z}$ such that x is positive and is not equal to the sum of three integer squares.

The symbol $\exists$ is called the *existential quantifier*. To prove that an existence statement is true, it is enough to find just one object satisfying the required property. So (1) is true, since $n = -3$ has the required property; and (3) is true since $x = 7$ is not the sum of three squares (of course there are many other values of x having this property, but only one value is required to demonstrate the truth of (3)).

Now consider the following statements:

(4) For all integers n, $n^2 \geq 0$.

(5) The cube of any integer is positive.

(6) Every integer is equal to the difference of two positive integers.

All these statements are of the form: "for all integers, a certain property is true." Again, this type of statement is very common in mathematics, and we represent the phrase "for all" by a special symbol, namely $\forall$. So the above statements can be rewritten as follows:

(4) $\forall n \in \mathbb{Z}$, $n^2 \geq 0$.

(5) $\forall n \in \mathbb{Z}$, $n^3 > 0$.

(6) $\forall x \in \mathbb{Z}$, x is equal to the difference of two positive integers.

The symbol $\forall$ is called the *universal quantifier*. To show that a "for all" state-ment is true, a general argument is required; to show it is false, a single coun-terexample is all that is needed (this is just proof by counterexample, discussed in the previous section). I will leave you to show that (4) and (6) are true, while (5) is false.

Many mathematical statements involve more than one quantifier. For exam-ple, statement (6) above can be rewritten as

(6) $\forall x \in \mathbb{Z}$, $\exists m, n \in \mathbb{Z}$ such that $m > 0, n > 0$ and $x = m - n$.

Here's another example: the statement "for any integer a, there is an integer b such that $a + b = 0$" can be rewritten as "$\forall a \in \mathbb{Z}, \exists b \in \mathbb{Z}$ such that $a + b = 0$." Notice that the order of quantifiers is important: the statement "$\exists b \in \mathbb{Z}$ such that $\forall a \in \mathbb{Z}$, $a + b = 0$" means something quite different.

Let's finish by seeing how to find the negation of a statement involving quan-tifiers. Consider statement (1) above: $\exists n \in \mathbb{Z}$ such that $n^3 = -27$. The negation of this is the statement "there does not exist an integer n such that $n^3 = -27$" — in other words, "every integer has cube not equal to -27," or more suc-cinctly, "$\forall n \in \mathbb{Z}, n^3 \neq -27$." So to form the negation of the original statement, we have changed $\exists$ to $\forall$ and negated the conclusion (i.e., changed $n^3 = -27$ to $n^3 \neq -27$).

Now consider statement (5): $\forall n \in \mathbb{Z}$, $n^3 > 0$. The negation of this is "not all integers have a positive cube" — in other words, "there is an integer having a non-positive cube," or more succinctly, "$\exists n \in \mathbb{Z}$ such that $n^3 \leq 0$." This time, to form the negation we have replaced $\forall$ by $\exists$ and negated the conclusion.

To summarise: when forming the negation of a statement involving quantifiers, we change $\exists$ to $\forall$, change $\forall$ to $\exists$ and negate the conclusion.

Let's do another example, and negate the following statement:

(7) For any integers x and y, there is an integer z such that $x^2 + y^2 = z^2$.

We can rewrite this as: $\forall x \in \mathbb{Z}, \forall y \in \mathbb{Z}$, $\exists z \in \mathbb{Z}$ such that $x^2 + y^2 = z^2$. Hence the negation is

$$\exists x \in \mathbb{Z}, \exists y \in \mathbb{Z}, \text{ such that } \forall z \in \mathbb{Z}, \; x^2 + y^2 \neq z^2.$$

In other words: there exist integers x, y such that for all integers z, $x^2 + y^2 \neq z^2$. I'm sure you can pretty quickly decide whether (7) or its negation is true.

Finally, let me make an observation for you to be wary of or amused by (or both). Here are a couple of strange statements involving the empty set:

(8) $\forall a \in \{x \mid x \text{ a real number}, x^2 + 1 = 0\}$, we have $a^{17} - 72a^{12} + 39 = 0$.

(9) $\exists b \in \{x \mid x \text{ a real number}, x^2 + 1 = 0\}$ such that $b^2 \geq 0$.

You will have noticed that the set $\{x \mid x \text{ a real number}, x^2 + 1 = 0\}$ is equal to the empty set. Hence the statement in (8) says that all elements of the empty set have a certain property; this is true, since there are no elements in the empty set! Likewise, any similar "for all" statement involving the empty set is true. On the other hand, the statement (9) says that there exists an element of the empty set with a certain property; this must be false, since there are no elements in the empty set.

Exercises for Chapter 1

1. Let A be the set $\{\alpha, \{1, \alpha\}, \{3\}, \{\{1, 3\}\}, 3\}$. Which of the following statements are true and which are false?

    (a) $\alpha \in A$.
    (b) $\{\alpha\} \notin A$.
    (c) $\{1, \alpha\} \subseteq A$.
    (d) $\{3, \{3\}\} \subseteq A$.
    (e) $\{1, 3\} \in A$.

    (f) $\{\{1, 3\}\} \subseteq A$.
    (g) $\{\{1, \alpha\}\} \subseteq A$.
    (h) $\{1, \alpha\} \notin A$.
    (i) $\emptyset \subseteq A$.

2. Let B, C, D, E be the following sets:

$$B = \{x \mid x \text{ a real number, } x^2 < 4\},$$

$$C = \{x \mid x \text{ a real number, } 0 \le x < 2\},$$

$$D = \{x \mid x \in \mathbb{Z}, x^2 < 1\},$$

$$E = \{1\}.$$

(a) Which pair of these sets has the property that neither is contained in the other?

(b) You are given that X is one of the sets B, C, D, E, but you do not know which one. You are also given that $E \subseteq X$ and $X \subseteq B$. What can you deduce about X?

3. Which of the following arguments are valid? For the valid ones, write down the argument symbolically.

(a) I eat chocolate if I am depressed. I am not depressed. Therefore I am not eating chocolate.

(b) I eat chocolate only if I am depressed. I am not depressed. Therefore I am not eating chocolate.

(c) If a movie is not worth seeing, then it was not made in England. A movie is worth seeing only if critic Ivor Smallbrain reviews it. The movie *Cat on a Hot Tin Proof* was not reviewed by Ivor Smallbrain. Therefore *Cat on a Hot Tin Proof* was not made in England.

4. A and B are two statements. Which of the following statements about A and B implies one or more of the other statements?

(a) Either A is true or B is true.

(b) $A \Rightarrow B$.

(c) $B \Rightarrow A$.

(d) $\bar{A} \Rightarrow B$.

(e) $\bar{B} \Rightarrow A$.

5. Which of the following statements are true, and which are false?

(a) $n = 3$ only if $n^2 - 2n - 3 = 0$.

(b) $n^2 - 2n - 3 = 0$ only if $n = 3$.

(c) If $n^2 - 2n - 3 = 0$ then $n = 3$.

(d) For integers a and b, ab is a square only if both a and b are squares.

(e) For integers a and b, ab is a square if both a and b are squares.

6. Write down careful proofs of the following statements:

    (a) $\sqrt{6} - \sqrt{2} > 1$.

    (b) If n is an integer such that n^2 is even, then n is even.

    (c) If $n = m^3 - m$ for some integer m, then n is a multiple of 6.

7. Disprove the following statements:

    (a) If n and k are positive integers, then $n^k - n$ is always divisible by k.

    (b) Every positive integer is the sum of three squares (the squares being 0, 1, 4, 9, 16, etc.).

8. Given that the number 8881 is not a prime number, prove that it has a prime factor that is at most 89. (*Hint: Don't try to factorize 8881! Try to be a bit more clever and prove it by contradiction.*)

9. In this question I am assuming you know what a prime number is; if not, take a look at the definition on page 69.

    For each of the following statements, form its negation and either prove that the statement is true or prove that its negation is true:

    (a) $\forall n \in \mathbb{Z}$ such that n is a prime number, n is odd.

    (b) $\forall n \in \mathbb{Z}$, $\exists a,b,c,d,e,f,g,h \in \mathbb{Z}$ such that

    $$n = a^3 + b^3 + c^3 + d^3 + e^3 + f^3 + g^3 + h^3.$$

    (c) $\exists x \in \mathbb{Z}$ such that $\forall n \in \mathbb{Z}$, $x \neq n^2 + 2$.

    (d) $\exists x \in \mathbb{Z}$ such that $\forall n \in \mathbb{Z}$, $x \neq n + 2$.

    (e) $\forall y \in \{x \mid x \in \mathbb{Z}, x \geq 1\}$, $5y^2 + 5y + 1$ is a prime number.

    (f) $\forall y \in \{x \mid x \in \mathbb{Z}, x^2 < 0\}$, $5y^2 + 5y + 1$ is a prime number.

10. Prove by contradiction that a real number that is less than every positive real number cannot be positive.

11. Critic Ivor Smallbrain (see Exercise 3(c)) has been keeping a careful account of the number of chocolate bars he has eaten during film screenings over his career. For each positive integer n he denotes by a_n the total number of bars he consumed during the first n films. One evening, during a screening of the Christmas epic *It's a Wonderful Proof*, he notices that the sequence $a_1, a_2, a_3, \ldots, a_n, \ldots$ obeys the following rules for all $n \geq 1$:

    $$a_{n+1} > a_n, \text{ and } a_{a_n} = 3n.$$

    Also $a_1 > 0$.

(a) Find a_1. (*Hint:* Let $x = a_1$. Then what is a_x?)

(b) Find $a_2, a_3, \ldots, a_9$.

(c) Find a_{100}.

(d) Investigate the sequence $a_1, a_2, \ldots, a_n, \ldots$ further.

Chapter 2

Number Systems

In this chapter we introduce three number systems: the real numbers, the integers and the rationals.

The Real Numbers

Here is an infinite straight line:

Choose a point on this line and label it as 0. Also choose a unit of length, and use it to mark off evenly spaced points on the line, labelled by the whole numbers $\ldots, -2, -1, 0, 1, 2, \ldots$ like this:

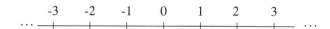

We shall think of the real numbers as the points on this line. Viewed in this way, the line is called the *real line*. Write $\mathbb{R}$ for the set of all real numbers.

The real numbers have a natural *ordering*, which we now describe. If x and y are real numbers, we write $x < y$, or equivalently $y > x$, if x is to the left of y on the real line; under these circumstances we say x is less than y, or y is greater than x. Also, $x \leq y$ indicates that x is less than or equal to y. Thus, the following statements are all true: $1 \leq 1, 1 \geq 1, 1 < 2, 2 \geq 1$. A real number x is *positive* if $x > 0$ and is *negative* if $x < 0$.

The *integers* are the whole numbers, marked as above on the real line. We write $\mathbb{Z}$ for the set of all integers and $\mathbb{N}$ for the set of all positive integers $\{1, 2, 3, \ldots\}$. Positive integers are sometimes called *natural numbers*.

Fractions $\frac{m}{n}$ can also be marked on the real line. For example, $\frac{1}{2}$ is placed halfway between 0 and 1; in general, $\frac{m}{n}$ can be marked by dividing each of the unit intervals into n equal sections and counting m of these sections away from 0. A real number of the form $\frac{m}{n}$ (where m, n are integers) is called a *rational number*. We write $\mathbb{Q}$ for the set of all rational numbers.

There are of course many different fractions representing the same rational number: for example, $\frac{8}{12} = \frac{-6}{-9} = \frac{2}{3}$, and so on. We say the rational $\frac{m}{n}$ is in *lowest terms* if no cancelling is possible — that is, if m and n have no common factors (apart from 1 and -1).

Rationals can be added and multiplied according to the familiar rules:

$$\frac{m}{n} + \frac{p}{q} = \frac{mq + np}{nq}, \quad \frac{m}{n} \times \frac{p}{q} = \frac{mp}{nq}.$$

Notice that the sum and product of two rationals is again rational.

In fact, addition and multiplication of arbitrary real numbers can be defined in such a way as to obey the following rules:

RULES 2.1   *For all $a, b, c, \in \mathbb{R}$,*
(1) $a + b = b + a$ and $ab = ba$
(2) $a + (b + c) = (a + b) + c$ and $a(bc) = (ab)c$
(3) $a(b + c) = ab + ac$.

For example, (2) assures us that $(2 + 5) + (-3) = 2 + (5 + (-3))$ (i.e., $7 - 3 = 2 + 2$), and $(2 \times 5) \times (-3) = 2 \times (5 \times (-3))$ (i.e., $10 \times (-3) = 2 \times (-15)$).

Before proceeding, let us pause briefly to reflect on these rules. They may seem "obvious" in some sense, in that you have probably been assuming them for years without thinking. But ponder the following equation, to be solved for x:

$$x + 3 = 5.$$

What are the steps we carry out when we solve this equation? Here they are:
   Step 1. Add -3 to both sides: $(x + 3) + (-3) = 5 + (-3)$.
   Step 2. Apply rule (2): $x + (3 + (-3)) = 5 - 3$.
   Step 3. This gives $x + 0 = 5 - 3$, hence $x = 2$.
The point is that without rule (2) we would be stuck. (Indeed, there are strange systems of objects with an addition for which one does not have rule (2), and in such systems one cannot even solve simple equations like the one above.)

There are some further important rules obeyed by the real numbers, relating to the ordering described above. We postpone discussion of these until Chapter 5.

Rationals and Irrationals

We often call a rational number simply a rational. The next result shows that the rationals are densely packed on the real line.

PROPOSITION 2.1

Between any two rationals there is another rational.

PROOF   Let r and s be two different rationals. Say r is the larger, so $r > s$. We claim that the real number $\frac{1}{2}(r+s)$ is a rational lying between r and s. To see this, observe that $\frac{1}{2}r > \frac{1}{2}s \Rightarrow \frac{1}{2}r + \frac{1}{2}s > \frac{1}{2}s + \frac{1}{2}s \Rightarrow \frac{1}{2}(r+s) > s$, and likewise $\frac{1}{2}r > \frac{1}{2}s \Rightarrow \frac{1}{2}r + \frac{1}{2}r > \frac{1}{2}s + \frac{1}{2}r \Rightarrow r > \frac{1}{2}(r+s)$. Thus, $\frac{1}{2}(r+s)$ lies between r and s. Finally, it is rational, since if $r = \frac{m}{n}, s = \frac{p}{q}$, then $\frac{1}{2}(r+s) = \frac{mq+np}{2nq}$. ∎

Despite its innocent statement and quick proof, this is a rather significant result. For example, it implies that in contrast to the integers, there is no smallest positive rational, since for any positive rational x there is a smaller positive rational (for example, $\frac{1}{2}x$); likewise, given any rational, there is no "next rational up." The proposition also shows that the rationals cannot be represented completely by "dots" on the real line, since between any two dots there would have to be another dot.

The proposition also raises a profound question: OK, the rationals are dense on the real line; but do they in fact fill out the whole line? In other words, is every real number a rational?

The answer is no, as we shall now demonstrate. First we need the following proposition, which is not quite as obvious as it looks.

PROPOSITION 2.2

There is a real number α such that $\alpha^2 = 2$.

PROOF   Draw a square of side 1:

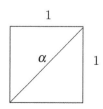

Let α be the length of a diagonal of the square. Then by Pythagoras, $\alpha^2 = 2$.   ∎

For the real number α in Proposition 2.2, we adopt the usual notation $\alpha = \sqrt{2}$.

PROPOSITION 2.3
$\sqrt{2}$ *is not rational.*

PROOF   This is a proof by contradiction. Suppose the statement is false — that is, suppose $\sqrt{2}$ is rational. This means that there are integers m, n such that

$$\sqrt{2} = \frac{m}{n}.$$

Take $\frac{m}{n}$ to be in lowest terms (recall that this means that m, n have no common factors greater than 1).

Squaring the above equation gives $2 = \frac{m^2}{n^2}$, hence

$$m^2 = 2n^2.$$

If m was odd, then m^2 would be odd (by Example 1.2); but $m^2 = 2n^2$ is clearly even, so this cannot be the case. Therefore, m is even. Hence, we can write $m = 2k$, where k is an integer. Then

$$m^2 = 4k^2 = 2n^2.$$

Consequently $n^2 = 2k^2$. So n^2 is even and, again by Example 1.2, this means n is also even.

We have now shown that both m and n are even. However, this means that the fraction $\frac{m}{n}$ is *not* in lowest terms. This is a contradiction. There-fore, $\sqrt{2}$ is not rational.   ∎

The following slightly more complicated geometrical argument than that given in Proposition 2.2 shows the existence of the real number $\sqrt{n}$ for any positive integer n. As in the figure on the next page, draw a circle with diameter AB, with a point D marked so that $AD = n, DB = 1$. We leave it to the reader to use Pythagoras in the right-angled triangles ACD, BCD and ABC to show that the length CD has square equal to n, and hence $CD = \sqrt{n}$.

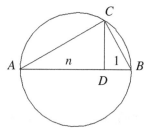

A real number that is not a rational is called an *irrational* number (or just an irrational). Thus $\sqrt{2}$ is an irrational, by Proposition 2.3. The next result enables us to construct many more examples of irrationals.

PROPOSITION 2.4
Let a be a rational number, and b an irrational.
(i) Then $a+b$ is irrational.
(ii) If $a \neq 0$, then ab is also irrational.

PROOF   (i) We prove this by contradiction. Suppose $a+b$ is rational, say $a+b = \frac{m}{n}$. Then writing the rational a as $\frac{p}{q}$, we have

$$b = \frac{m}{n} - a = \frac{m}{n} - \frac{p}{q}.$$

However, the right-hand side is rational, whereas b is given to be irrational, so this is a contradiction. Hence, $a+b$ is irrational.

The proof of part (ii) is very similar to that of (i), and we leave it to the reader. ∎

Example 2.1
The proposition shows that, for example, $1+\sqrt{2}$ and $-5\sqrt{2}$ are irrational; indeed, $r+s\sqrt{2}$ is irrational for any rationals r,s with $s \neq 0$. Note also that there exist many further irrationals, not of this form. For instance, $\sqrt{3}$ is irrational, and there are no rationals r,s such that $\sqrt{3} = r+s\sqrt{2}$ (see Exercise 1 on the next page).

Thus, there are "many" irrationals, in some sense. The next result shows that, like the rationals, the irrationals are densely packed on the real line.

PROPOSITION 2.5
Between any two real numbers there is an irrational.

PROOF   Let a and b be two real numbers, and say a is the smaller, so $a < b$. Choose a positive integer n that is larger than the real number $\frac{\sqrt{2}}{b-a}$. Then

$$\frac{\sqrt{2}}{n} < b - a.$$

If a is rational, then by Proposition 2.4, $a + \frac{\sqrt{2}}{n}$ is irrational; it also lies between a and b. And if a is irrational, then $a + \frac{1}{n}$ is irrational and lies between a and b.   ∎

Exercises for Chapter 2

1. (a) Prove that $\sqrt{3}$ is irrational. (*Hint:* Example 1.3 should be useful.)

   (b) Prove that there are no rationals r, s such that $\sqrt{3} = r + s\sqrt{2}$.

2. Which of the following numbers are rational and which are irrational ?

   (a) $\sqrt{2} + \sqrt{\frac{3}{2}}$.

   (b) $1 + \sqrt{2} + \sqrt{\frac{3}{2}}$.

   (c) $2\sqrt{18} - 3\sqrt{8} + \sqrt{4}$.

   (d) $\sqrt{2} + \sqrt{3} + \sqrt{5}$ (*Hint:* For this part you can use the fact that if n is a positive integer that is not a square, then $\sqrt{n}$ is irrational — we will prove this later, in Proposition 11.4.)

   (e) $\sqrt{2} + \sqrt{3} - \sqrt{5 + 2\sqrt{6}}$.

3. For each of the following statements, either prove it is true or give a counterexample to show it is false.

   (a) The product of two rational numbers is always rational.

   (b) The product of two irrational numbers is always irrational.

   (c) The product of two irrational numbers is always rational.

   (d) The product of a non-zero rational and an irrational is always irrational.

4. (a) Let a, b be rationals and x irrational. Show that if $\frac{x+a}{x+b}$ is rational, then $a = b$.

   (b) Let x, y be rationals such that $\frac{x^2 + x + \sqrt{2}}{y^2 + y + \sqrt{2}}$ is also rational. Prove that either $x = y$ or $x + y = -1$.

5. Prove that if n is any positive integer, then $\sqrt{n} + \sqrt{2}$ is irrational.

6. Prove that between any two different real numbers there is a rational number and an irrational number.

7. Find n, given that both n and $\sqrt{n-2} + \sqrt{n+2}$ are positive integers. (*Hint:* The gaps between squares n^2, $(n+1)^2, \ldots$ grow as n increases.)

8. Critic Ivor Smallbrain is watching the horror movie *Salamanders on a Desert Island*. In the film, there are 30 salamanders living on a desert island: 15 are red, 7 blue and 8 green. When two of a different colour meet, horrifyingly they both change into the third colour. (For example, if a red and a green meet, they both become blue.) When two of the same colour meet, they change into both of the other colours. (For example, if two reds meet, one becomes green and one becomes blue.) It is all quite terrifying.

In between being horrified and terrified, Ivor idly wonders whether it could ever happen that at some instant in the future, all of the salamanders would be red. Can you help him ? (*Hint:* Consider the remainders of the totals of each colour when you divide by 3.)

Chapter 3

Decimals

It is all very well to have the real number system as points on the real line, but it is hard to prove any interesting facts about the reals without any convenient notation for them. We now remedy this by introducing the decimal notation for reals and demonstrating a few of its basic properties.

We are all familiar with the following decimal expressions:

$$\frac{1}{2} = 0.50000\ldots$$

$$\frac{1}{9} = 0.11111\ldots$$

$$\frac{1}{7} = 0.142857142857\ldots$$

But what do we mean when we write, for example, $\frac{1}{9} = 0.11111\ldots$? We mean that the "sum to infinity" of the series

$$0.1111\ldots = 0.1 + 0.01 + 0.001 + \cdots = \frac{1}{10} + \frac{1}{10^2} + \frac{1}{10^3} + \cdots$$

is $\frac{1}{9}$; in other words, we can get as close as we like to $\frac{1}{9}$ provided we sum enough terms of the series. To make this absolutely precise would require us to go into the concepts of limits and convergence, which we shall do much later in Chapter 23. For now, I hope the meaning is reasonably clear.

The above fact about $\frac{1}{9}$ is a special case of the following result on geometric series, which is probably very familiar.

PROPOSITION 3.1

Let x be a real number.
(i) If $x \neq 1$, then $x + x^2 + x^3 + \cdots + x^n = \frac{x(1-x^n)}{1-x}$.
(ii) If $-1 < x < 1$, then the sum to infinity

$$x + x^2 + x^3 + \cdots = \frac{x}{1-x}.$$

PROOF   (i) Let $s_n = x + x^2 + x^3 + \cdots + x^n$. Then $xs_n = x^2 + x^3 + \cdots + x^n + x^{n+1}$. Subtracting, we get $(1-x)s_n = x - x^{n+1}$, which gives (i).

(ii) Since $-1 < x < 1$, we can make x^n as small as we like, provided we take n large enough. So we can make the sum in (i) as close as we like to $\frac{x}{1-x}$ provided we sum enough terms. That is to say, the sum to infinity is $\frac{x}{1-x}$.   ∎

Putting $x = \frac{1}{10}$ in this proposition gives $\frac{1}{10} + \frac{1}{10^2} + \frac{1}{10^3} + \cdots = \frac{1}{9}$, as claimed above.

Likewise, in general, the decimal expression $a_0.a_1a_2a_3\ldots$ where a_0 is an integer and $a_1, a_2, \ldots$ are integers between 0 and 9, means the real number that is the sum to infinity of the series

$$a_0 + \frac{a_1}{10} + \frac{a_2}{10^2} + \frac{a_3}{10^3} + \cdots$$

With this understanding, we obtain the next result, which gives us the convenient decimal notation for all real numbers.

PROPOSITION 3.2
Every real number x has a decimal expression

$$x = a_0.a_1a_2a_3\ldots.$$

PROOF   Picture x on the real line. Certainly x lies between two consecutive integers; let a_0 be the lower of these, so that

$$a_0 \leq x < a_0 + 1.$$

Now divide the line between a_0 and $a_0 + 1$ into ten equal sections. Certainly x lies in one of these sections, so we can find a_1 between 0 and 9 such that

$$a_0 + \frac{a_1}{10} \leq x < a_0 + \frac{a_1 + 1}{10}.$$

Similarly, we can find a_2 such that

$$a_0 + \frac{a_1}{10} + \frac{a_2}{10^2} \leq x < a_0 + \frac{a_1}{10} + \frac{a_2 + 1}{10^2},$$

and so on. If we do this enough times, the sum $a_0 + \frac{a_1}{10} + \frac{a_2}{10^2} + \cdots$ gets as close as we like to x. As explained above, this is what we mean by saying that $x = a_0.a_1a_2a_3\ldots$.   ∎

Example 3.1
We use the method of the proof just given to find the first few digits in the decimal expression for $\sqrt{2}$. Let $\sqrt{2} = a_0.a_1a_2a_3\ldots$. First, observe

that $1^2 = 1$ and $2^2 = 4$, so $\sqrt{2}$ lies between 1 and 2, and hence $a_0 = 1$. Next, $(1.4)^2 = 1.96$, while $(1.5)^2 = 2.25$, so $a_1 = 4$. Likewise, $(1.41)^2 < 2$ while $(1.42)^2 > 2$, so $a_2 = 1$. We can continue finding decimal digits in this way until we get really fed up.

We now have a convenient notation for all real numbers: they all have decimal expressions. Two basic questions about this notation arise immediately:

(1) Can the same real number have two different decimal expressions; and if so, can we describe exactly when this happens?
(2) Which decimal expressions are rational and which are irrational?

We shall answer these questions in the next few results.

For (1), notice first that

$$0.9999\ldots = \frac{9}{10} + \frac{9}{10^2} + \frac{9}{10^3} + \cdots$$

$$= 9\left(\frac{1}{10} + \frac{1}{10^2} + \frac{1}{10^3} + \cdots\right) = 9\left(\frac{1}{9}\right) = 1.$$

Thus, the real number 1 has two different decimal expressions:

$$1 = 1.0000\ldots = 0.9999\ldots$$

Similarly, for example,

$$0.2579999\ldots = 0.2580000\ldots, \quad \text{and} \quad 1299.9999\ldots = 1300.0000\ldots,$$

and so on. Is this the only way two different decimal expressions can be equal? The answer is yes.

PROPOSITION 3.3
Suppose that $a_0.a_1a_2a_3\ldots$ and $b_0.b_1b_2b_3\ldots$ are two different decimal expressions for the same real number. Then one of these expressions ends in $9999\ldots$ and the other ends in $0000\ldots$.

PROOF Suppose first that $a_0 = b_0 = 0$. Call the real number with these two expressions x, so that

$$x = 0.a_1a_2a_3\ldots = 0.b_1b_2b_3\ldots \tag{3.1}$$

Let the first place where the two expressions disagree be the kth place (k could be 1 of course). Thus $x = 0.a_1\ldots a_{k-1}a_k\ldots = 0.a_1\ldots a_{k-1}b_k\ldots$, where $a_k \neq b_k$. There is no harm in assuming $a_k > b_k$, hence $a_k \geq b_k + 1$. Then

$$x \geq 0.a_1\ldots a_{k-1}a_k000\ldots$$

and
$$x \leq 0.a_1 \ldots a_{k-1} b_k 999 \ldots = 0.a_1 \ldots a_{k-1}(b_k + 1)000 \ldots.$$

It follows that $a_k = b_k + 1$ and that the two expressions for x in (3.1) are $0.a_1 \ldots a_k 000 \ldots$ and $0.a_1 \ldots a_{k-1}(a_k - 1)999 \ldots.$

Finally, to handle the general case (where a_0, b_0 are not assumed to be 0), we replace a_0, b_0 with their expressions as integers using decimal digits and apply the above argument. ∎

This provides us with a satisfactory answer to our question (1) above.

Now we address question (2): Which decimal expressions are rational, and which are irrational? Choose a couple of rationals at random — say $\frac{8}{7}$ and $\frac{13}{22}$ — and work out their decimal expressions:

$$\frac{8}{7} = 1.142857142857\ldots, \quad \frac{13}{22} = 0.59090909\ldots.$$

We observe that they have a striking feature in common: there is a sequence of digits that eventually repeats forever. We call such a decimal expression *periodic*.

In general, a periodic decimal is one that takes the form

$$a_0.a_1 \ldots a_k \, b_1 \ldots b_l \, b_1 \ldots b_l \, b_1 \ldots b_l \ldots.$$

We abbreviate this expression by writing it as $a_0.a_1 \ldots a_k \overline{b_1 \ldots b_l}$. The *period* of such a decimal is the number of digits in a repeating sequence of smallest length. For example, the decimal expression for $\frac{8}{7}$ has period 6.

The next result should not come as a major surprise.

PROPOSITION 3.4
The decimal expression for any rational number is periodic.

PROOF   Consider a rational $\frac{m}{n}$ (where $m, n \in \mathbb{Z}$). To express this as a decimal, we perform long division of n into $m.0000\ldots$. At each stage of the long division, we get a remainder which is one of the n integers between 0 and $n-1$. Therefore, eventually we must get a remainder that occurred before. The digits between the occurrences of these remainders will then repeat forever. ∎

Proposition 3.4 tells us that

$$a_0.a_1 a_2 a_3 \ldots \text{ rational} \Rightarrow a_0.a_1 a_2 a_3 \ldots \text{ periodic.}$$

It would be very nice if the reverse implication were also true — that is, periodic $\Rightarrow$ rational. Let us first consider an example.

Example 3.2

Let $x = 0.3\overline{14}$. Is x rational? Well,

$$x = \frac{3}{10} + \frac{14}{10^3} + \frac{14}{10^5} + \frac{14}{10^7} + \cdots = \frac{3}{10} + \frac{14}{10^3}\left(1 + \frac{1}{10^2} + \frac{1}{10^4} + \cdots\right).$$

The series in the parentheses is a geometric series, which by Proposition 3.1 has sum to infinity $\frac{100}{99}$, so

$$x = \frac{3}{10} + \frac{14}{10^3} \cdot \frac{100}{99} = \frac{311}{990}.$$

In particular, x is rational.

It is not at all hard to generalize this argument to show that the reverse implication (periodic $\Rightarrow$ rational) is indeed true:

PROPOSITION 3.5

Every periodic decimal is rational.

PROOF Let $x = a_0.a_1 \ldots a_k \overline{b_1 \ldots b_l}$ be a periodic decimal. Define

$$A = a_0.a_1 \ldots a_k, \quad B = 0.b_1 \ldots b_l.$$

Then A and B are both rationals, and

$$x = A + \frac{B}{10^k}\left(1 + \frac{1}{10^l} + \frac{1}{10^{2l}} + \cdots\right) = A + \frac{B}{10^k} \cdot \frac{10^l}{10^l - 1},$$

which is clearly also rational. ∎

Exercises for Chapter 3

1. Express the decimal $1.\overline{813}$ as a fraction $\frac{m}{n}$ (where m and n are integers).

2. Show that the decimal expression for $\sqrt{2}$ is not periodic.

3. Which of the following numbers are rational, and which are irrational? Express those which are rational in the form $\frac{m}{n}$ with $m, n \in \mathbb{Z}$.

   (a) $0.a_1 a_2 a_3 \ldots$, where for $n = 1, 2, 3, \ldots$, the value of a_n is the number $0,1,2,3$ or 4 which is the remainder on dividing n by 5.

(b) $0.101001000100001000001\ldots$

(c) $1.b_1b_2b_3\ldots$, where $b_i = 1$ if i is a square, and $b_i = 0$ if i is not a square.

4. Without using a calculator, find the cube root of 2, correct to 1 decimal place.

5. The *Fibonacci sequence* starts with the terms $1,1$ and then proceeds by letting the next term be the sum of the previous two terms. So the sequence starts $1,1,2,3,5,8,13,21,34,\ldots$. With this in mind, consider the decimal expansion of $x = \frac{100}{9899}$: it is $0.0101020305081321341\ldots$. Note how the Fibonacci sequence lives inside this expansion. Can you explain this? Do you think it continues forever?

(*Hint:* First show that $100 + x + 100x = 10000x$.)

6. Show that for an integer $n \geq 2$, the period of the decimal expression for the rational number $\frac{1}{n}$ is at most $n - 1$.

Find the first few values of n for which the period of $\frac{1}{n}$ is equal to $n - 1$. Do you notice anything interesting about the values you've found?

7. Prove that a rational $\frac{m}{n}$ (in lowest terms) has a decimal expression ending in repeating zeroes, if and only if the denominator n is of the form 2^a5^b, where $a, b \geq 0$ and a, b are integers.

8. Critic Ivor Smallbrain is watching the classic film *11.9 Angry Men*. But he is bored, and starts wondering idly exactly which rational numbers $\frac{1}{n}$ have decimal expressions with period equal to 1. Having done the previous question during the advertisments before the film, he notices that the period is 1 if the denominator n is 2^a5^b. But he also notices some other values of n for which the period is 1, such as $n = 3$.

Can you help Ivor and find all the values of n for which $\frac{1}{n}$ has period 1?

Chapter 4

n^{th} Roots and Rational Powers

In Chapter 2, just after proving Proposition 2.3, we gave a cunning geometrical construction that demonstrated the existence of the real number $\sqrt{n}$ for any positive integer n. However, proving the existence of a cube root and, more generally, an nth root of any positive real number x is much harder and requires a deeper analysis of the reals than we have undertaken thus far. We shall carry out such an analysis later, in Chapter 24. However, because we wish to include n^{th} roots in the discussion of complex numbers in the next chapter, we pick out the main result from Chapter 24 on such matters, namely Proposition 24.2, and state it here. (It is, of course, proved in Chapter 24.)

PROPOSITION 4.1
Let n be a positive integer. If x is a positive real number, then there is exactly one positive real number y such that $y^n = x$.

If x, y are as in the statement, we adopt the familiar notation

$$y = x^{\frac{1}{n}}.$$

Thus, for example, $5^{\frac{1}{2}}$ is the positive square root of 5, and $5^{\frac{1}{7}}$ is the unique positive real number y such that $y^7 = 5$.

We can extend this notation to define rational powers of positive reals as follows. Let $x > 0$. Integer powers x^m ($m \in \mathbb{Z}$) are defined in the familiar way: if $m > 0$ then $x^m = xx\ldots x$, the product of m copies of x, and $x^{-m} = \frac{1}{x^m}$; and for $m = 0$ we define $x^0 = 1$.

Now let $\frac{m}{n} \in \mathbb{Q}$ (with $m, n \in \mathbb{Z}$ and $n \geq 1$). Then we define

$$x^{\frac{m}{n}} = \left(x^{\frac{1}{n}}\right)^m.$$

For example, $5^{-\frac{4}{7}}$ is defined to be $(5^{\frac{1}{7}})^{-4}$.

The basic rules concerning products of these rational powers are given in the next proposition. Although they probably seem rather familiar, they are not totally obvious and require careful proof.

PROPOSITION 4.2
Let x, y be positive real numbers and $p, q \in \mathbb{Q}$. Then

(i) $x^p x^q = x^{p+q}$

(ii) $(x^p)^q = x^{pq}$

(iii) $(xy)^p = x^p y^p$

PROOF   (i) We first establish the result when p and q are both integers. In this case, when $p, q \geq 0$, we have $x^p = x \ldots x$ (p factors), $x^q = x \ldots x$ (q factors), so

$$x^p x^q = (x \ldots x).(x \ldots x) = x^{p+q},$$

and when $p \geq 0, q < 0$, $x^q = 1/x \ldots x$ ($-q$ factors), so

$$x^p x^q = (x \ldots x)/(x \ldots x) = x^{p-(-q)} = x^{p+q}.$$

Similar arguments cover the other possibilities $p < 0, q \geq 0$ and $p, q < 0$.

Now let us consider the general case, where p, q are rationals. Write $p = \frac{m}{n}, q = \frac{h}{k}$ with $m, n, h, k \in \mathbb{Z}$. Then

$$x^p x^q = x^{\frac{m}{n}} x^{\frac{h}{k}} = x^{\frac{mk}{nk}} x^{\frac{hn}{nk}} = \left(x^{\frac{1}{nk}} \right)^{mk} \left(x^{\frac{1}{nk}} \right)^{hn}.$$

By the integer case of part (i), established in the previous paragraph, this is equal to

$$\left(x^{\frac{1}{nk}} \right)^{mk+hn},$$

which, by our definition of rational powers, is equal to

$$x^{\frac{mk+hn}{nk}} = x^{\frac{m}{n}+\frac{h}{k}} = x^{p+q}.$$

(ii, iii) First, as in (i), we easily establish the results for $p, q \in \mathbb{Z}$. Now let $p = \frac{m}{n}, q = \frac{h}{k}$ (with $h, k, m, n \in \mathbb{Z}$). By Proposition 4.1, there is a real number a such that $x = a^{nk}$. Then

$$(x^p)^q = \left(\left(a^{nk} \right)^{\frac{m}{n}} \right)^{\frac{h}{k}} = \left(\left(\left((a^k)^n \right)^{\frac{1}{n}} \right)^m \right)^{\frac{h}{k}} = \left(\left(a^k \right)^m \right)^{\frac{h}{k}} = \left(\left((a^m)^k \right)^{\frac{1}{k}} \right)^h$$

$$= (a^m)^h = a^{mh} = \left(\left(x^{1/nk} \right)^{mh} \right) = x^{\frac{mh}{nk}} = x^{pq}$$

and

$$x^p y^p = \left(x^{\frac{1}{n}}\right)^m \left(y^{\frac{1}{n}}\right)^m = \left(x^{\frac{1}{n}} y^{\frac{1}{n}}\right)^m = \left(\left(\left(x^{\frac{1}{n}} y^{\frac{1}{n}}\right)^n\right)^{\frac{1}{n}}\right)^m$$

$$= \left(\left(x^{\frac{1}{n}}\right)^n \left(y^{\frac{1}{n}}\right)^n\right)^{\frac{m}{n}} = (xy)^p. \quad \blacksquare$$

Exercises for Chapter 4

1. Show that $(50)^{3/4}\left(\frac{5}{\sqrt{2}}\right)^{-1/2} = 10$.

2. Simplify $2^{1/2} 5^{1/2} 4^{-1/4} 20^{1/4} 5^{-1/4} \sqrt{10}$.

3. What is the square root of 2^{1234}?

   What is the real cube root of $3^{(3^{333})}$?

4. Find an integer n and a rational t such that $n^t = 2^{1/2} 3^{1/3}$.

5. Which is bigger: 100^{10000} or 10000^{100}?

   Which is bigger: the cube root of 3 or the square root of 2? (No calculators allowed!)

6. Find all real solutions x of the equation $x^{1/2} - (2 - 2x)^{1/2} = 1$.

7. Prove that if $x, y > 0$ then $\frac{1}{2}(x+y) \geq \sqrt{xy}$. For which x, y does equality hold?

8. When we want to add three numbers, say $a + b + c$, we don't bother inserting parentheses because $(a + b) + c = a + (b + c)$. But with powers, this is not true — $(a^b)^c$ need not be equal to $a^{(b^c)}$ — so we must be careful. Show that this really is a problem, by finding positive integers a, b, c such that $(a^b)^c < a^{(b^c)}$ and positive integers d, e, f such that $(d^e)^f > d^{(e^f)}$.

9. After a delicious meal at the well-known French restaurant *La Racine et Puissance Rationelle*, critic Ivor Smallbrain notices that the bill comes to x pounds, y pence, where x and y are the smallest integers greater than 1 that satisfy the equation $y^{4/3} = x^{5/6}$.

   How much is the bill?

Chapter 5

Inequalities

An *inequality* is a statement about real numbers involving one of the symbols
">," "≥," "<" or "≤"; for example, $x > 2$ or $x^2 - 4y \leq 2x + 2$. In this chapter
we shall present some elementary notions concerning manipulation of inequalities.

Recall from Chapter 2 the basic Rules 2.1 satisfied by addition and multiplication of real numbers. As we mentioned there, there are various further rules concerning the ordering of the real numbers. Here they are:

RULES 5.1

(1) If $x \in \mathbb{R}$, then either $x > 0$ or $x < 0$ or $x = 0$ (and just one of these is true).

(2) If $x > y$ then $-x < -y$.

(3) If $x > y$ and $c \in \mathbb{R}$, then $x + c > y + c$.

(4) If $x > 0$ and $y > 0$, then $xy > 0$.

(5) If $x > y$ and $y > z$ then $x > z$.

Notice that rule (3) implies rule (2), since if $x > y$ then $x + (-x - y) > y + (-x - y)$ (taking $c = -x - y$ in rule (3)), which implies that $-y > -x$. However I have included rule (2) in the list above as it is useful to have it there explicitly.

The rest of the chapter consists of several examples showing how to use these rules to manipulate inequalities.

Example 5.1
If $x < 0$ then $-x > 0$.

PROOF   Applying (2) with 0 instead of x, and x instead of y, we see that $x < 0 \Rightarrow -x > 0$.   ∎

Example 5.2
If $x \neq 0$ then $x^2 > 0$.

PROOF   If $x > 0$ then by (4), $x^2 = xx > 0$. If $x < 0$ then $-x > 0$ (by Example 5.1), so (4) gives $(-x)(-x) > 0$; i.e., $x^2 > 0$.   ∎

Example 5.3
If $x > 0$ and $u > v$ then $xu > xv$.

PROOF   We have

$$
\begin{aligned}
u > v &\Rightarrow u - v > v - v = 0 \ \text{(by (3) with } c = -v) \\
&\Rightarrow x(u - v) > 0 \ \text{(by (4))} \\
&\Rightarrow xu - xv > 0 \\
&\Rightarrow xu - xv + xv > xv \ \text{(by (3) with } c = xv) \\
&\Rightarrow xu > xv. \quad ∎
\end{aligned}
$$

Example 5.4
If $u > v > 0$ then $u^2 > v^2$.

PROOF   Two applications of Example 5.3 give $u > v \Rightarrow u^2 > uv$ and $u > v \Rightarrow uv > v^2$. Hence $u^2 > v^2$ by (5).   ∎

Notice that reversing the roles of u and v, Example 5.4 also tells us that $u < v \Rightarrow u^2 < v^2$ (for positive u, v). Hence for positive u, v,

$$
u^2 \leq v^2 \Leftrightarrow u \leq v.
$$

Example 5.5
If $x > 0$ then $\frac{1}{x} > 0$.

PROOF   If $\frac{1}{x} < 0$ then $\frac{-1}{x} > 0$ by Example 5.1, so by (4), $x.\frac{-1}{x} > 0$ (i.e., $-1 > 0$), a contradiction. Therefore $\frac{1}{x} \geq 0$. Since $\frac{1}{x} \neq 0$, we conclude by Rule 5.1(1) that $\frac{1}{x} > 0$.   ∎

Example 5.6
Let $x_1, x_2, \ldots, x_n \in \mathbb{R}$, and suppose that k of these numbers are negative and the rest are positive. If k is even, then the product $x_1 x_2 \ldots x_n > 0$. And if k is odd, $x_1 x_2 \ldots x_n < 0$.

PROOF   Since the order of the x_is does not matter, we may as well assume that $x_1,\ldots,x_k$ are negative and $x_{k+1},\ldots,x_n$ are positive. Then by Example 5.1, $-x_1,\ldots,-x_k,x_{k+1},\ldots,x_n$ are all positive. By (4), the product of all of these is positive, so

$$(-1)^k x_1 x_2,\ldots,x_n > 0.$$

If k is even this says that $x_1 x_2,\ldots,x_n > 0$. And if k is odd it says that $-x_1 x_2,\ldots,x_n > 0$, hence $x_1 x_2,\ldots,x_n < 0$.   ∎

The next example is a typical elementary inequality to solve.

Example 5.7
For which values of x is $x < \frac{2}{x+1}$?

Answer First, a word of warning — we cannot multiply both sides by $x+1$, as this may or may not be positive. So we proceed more cautiously. Subtracting $\frac{2}{x+1}$ from both sides gives the inequality $x - \frac{2}{x+1} < 0$, which is the same as $\frac{x^2+x-2}{x+1} < 0$; that is,

$$\frac{(x+2)(x-1)}{x+1} < 0.$$

By Example 5.6, this is true if and only if either one or three of the quantities $x+2, x-1, x+1$ is negative. All three are negative when $x < -2$, and just one is negative when $-1 < x < 1$.

Example 5.8
Show that $x^2 + x + 1 > 0$ for all $x \in \mathbb{R}$.

Answer Note that $x^2 + x + 1 = (x+\frac{1}{2})^2 + \frac{3}{4}$. Hence, using Example 5.2 and Rule 5.1(3), we have $x^2 + x + 1 \geq \frac{3}{4}$ for all x.

For a real number x, we define the *modulus* of x, written $|x|$, by

$$|x| = \begin{cases} x, & \text{if } x \geq 0 \\ -x, & \text{if } x < 0 \end{cases}$$

For example, $|-5| = 5$ and $|7| = 7$.

Example 5.9
$|x|^2 = x^2$ and $|xy| = |x||y|$ for all $x, y \in \mathbb{R}$.

PROOF   If $x \geq 0$ then $|x| = x$ and $|x|^2 = x^2$; and if $x < 0$ then $|x| = -x$ and $|x|^2 = (-x)^2 = x^2$. The second part is Exercise 6 at the end of the chapter.   ∎

Notice that $|x|$ just measures the distance from the point x on the real line to the origin 0. Thus, for example, the set of values of x such that $|x| \leq 2$ consists of all x between -2 and 2, which we summarize as $-2 \leq x \leq 2$. More generally:

Example 5.10
Let $a, b \in \mathbb{R}$ with $b > 0$. For which values of x is the inequality $|x - a| \leq b$ satisfied?

Answer When $x \geq a$, the inequality says $x - a \leq b$; that is, $x \leq a + b$. And when $x < a$, the inequality says $a - x \leq b$; that is, $x \geq a - b$. So the range of values of x satisfying the inequality is $a - b \leq x \leq a + b$.

Example 5.11
Find all values of x such that $|x - 3| < 2|x + 3|$.

Answer We must be quite careful with this — the inequality varies according to whether $x < -3$, $-3 \leq x < 3$, or $x \geq 3$.

When $x < -3$ the inequality says $-(x - 3) < 2(-x - 3)$, which is the same as $x < -9$. When $-3 \leq x < 3$, the inequality says $-(x - 3) < 2(x + 3)$, which means that $3x > -3$, in other words $x > -1$. And when $x > 3$ the inequality says $x - 3 < 2(x + 3)$, which means that $x > -9$. We deduce that the values of x satisfying the inequality are

$$x < -9 \quad \text{and} \quad x > -1.$$

Example 5.12
The Triangle Inequality: $|x + y| \leq |x| + |y|$ for any $x, y \in \mathbb{R}$.

PROOF   By Example 5.9, we have

$$\begin{aligned}
|x + y|^2 &= (x + y)^2 \\
&= x^2 + 2xy + y^2 \\
&\leq |x|^2 + 2|x||y| + |y|^2 \\
&= (|x| + |y|)^2.
\end{aligned}$$

Hence $|x + y| \leq |x| + |y|$ by Example 5.4.   ∎

The following is an immediate consequence of the Triangle Inequality.

Example 5.13
$|x - y| \geq |x| - |y|$ for any $x, y \in \mathbb{R}$.

Example 5.14

If $x, y \in \mathbb{R}$ and $x \geq 0, y \geq 0$, then $\sqrt{xy} \leq \frac{1}{2}(x+y)$.

PROOF Using Example 5.4, we see that

$$
\begin{aligned}
\sqrt{xy} \leq \tfrac{1}{2}(x+y) &\Leftrightarrow xy \leq \tfrac{1}{4}(x+y)^2 \\
&\Leftrightarrow 4xy \leq x^2 + 2xy + y^2 \\
&\Leftrightarrow 0 \leq x^2 - 2xy + y^2 \\
&\Leftrightarrow 0 \leq (x-y)^2,
\end{aligned}
$$

which is true by Example 5.2. ∎

Although easy to prove, the bound in Example 5.14 is not as innocent as it looks. For example, it tells you that among all rectangles with a given area A, the square of side $\sqrt{A}$ has the smallest perimeter (see Exercise 8 at the end of the chapter). It is also the case $n = 2$ of a famous and very important inequality known as the "Arithmetic-Geometric Mean Inequality," which states that if n is a positive integer and $a_1, \ldots, a_n$ are positive real numbers, then

$$
(a_1 a_2 \cdots a_n)^{1/n} \leq \frac{1}{n}(a_1 + a_2 + \cdots + a_n). \tag{5.1}
$$

(The right hand side is the "arithmetic mean" of the numbers $a_1, \ldots, a_n$, and the left hand side is their "geometric mean.") We won't prove this inequality in this book, but in Exercise 9 you are asked to deduce it for some further special values of n.

Example 5.15

If a_1, a_2, b_1, b_2 are real numbers, then $a_1 b_1 + a_2 b_2 \leq \sqrt{a_1^2 + a_2^2}\sqrt{b_1^2 + b_2^2}$.

PROOF Observe that

$$
\begin{aligned}
a_1 b_1 + a_2 b_2 &\leq \sqrt{a_1^2 + a_2^2}\sqrt{b_1^2 + b_2^2} \\
&\Leftrightarrow (a_1 b_1 + a_2 b_2)^2 \leq (a_1^2 + a_2^2)(b_1^2 + b_2^2) \\
&\Leftrightarrow a_1^2 b_1^2 + 2a_1 b_1 a_2 b_2 + a_2^2 b_2^2 \leq a_1^2 b_1^2 + a_1^2 b_2^2 + a_2^2 b_1^2 + a_2^2 b_2^2 \\
&\Leftrightarrow 0 \leq a_1^2 b_2^2 - 2a_1 b_1 a_2 b_2 + a_2^2 b_1^2 \\
&\Leftrightarrow 0 \leq (a_1 b_2 - a_2 b_1)^2,
\end{aligned}
$$

which is true by Example 5.2. ∎

Example 5.15 is the case $n = 2$ of another famous inequality known as Cauchy's inequality; this one we *will* prove, in Chapter 8 (see Proposition 8.2).

Exercises for Chapter 5

1. Using Rules 5.1, show that if $x > 0$ and $y < 0$ then $xy < 0$, and that if $a > b > 0$ then $\frac{1}{a} < \frac{1}{b}$.

2. For which values of x is $x^2 + x + 1 \geq \frac{x-1}{2x-1}$?

3. For which values of x is $-3x^2 + 4x > 1$?

4. (a) Find the set of real numbers $x \neq 0$ such that $2x + \frac{1}{x} < 3$.

   (b) Find the set of real numbers t such that the equation $x^2 + tx + 3 = 0$ has two distinct real solutions.

5. Prove that if $0 < u < 1$ and $0 < v < 1$, then $\frac{u+v}{1+uv} < 1$. For which other values of u, v is this inequality true?

6. Prove that $|xy| = |x|\,|y|$ for all real numbers x, y.

7. Find the range of values of x such that

   (i) $|x+5| \geq 1$.

   (ii) $|x+5| > |x-2|$.

   (iii) $|x+5| < |x^2 + 2x + 3|$.

8. Prove the statement made after Example 5.14: among all rectangles with a given area A, the square of side $\sqrt{A}$ has the smallest perimeter.

9. By applying the inequality in Example 5.14 twice, prove that for any positive real numbers a_1, a_2, a_3, a_4,

$$(a_1 a_2 a_3 a_4)^{1/4} \leq \frac{1}{4}(a_1 + a_2 + a_3 + a_4).$$

   This is the Arithmetic-Geometric Mean Inequality (5.1) stated after Example 5.14, for the case $n = 4$. Try to deduce the case $n = 8$, and further cases.

10. Prove the following inequalities for any positive real numbers x, y:

    (i) $xy^3 \leq \frac{1}{4}x^4 + \frac{3}{4}y^4$

    (ii) $xy^3 + x^3 y \leq x^4 + y^4$.

11. Prove that if x, y, z are real numbers such that $x + y + z = 0$, then $xy + yz + zx \leq 0$.

12. Restless during a showing of the ten-hour epic *First Among Inequalities*, critic Ivor Smallbrain thinks of a new type of number, which he modestly decides to call "Smallbrain numbers." He calls an n-digit positive integer a Smallbrain number if it is equal to the sum of the nth powers of its digits. So for example, 371 is a Smallbrain number, since $371 = 3^3 + 7^3 + 1^3$.

Prove that there are no Smallbrain numbers with 1000 digits (i.e., there is no 1000-digit number that is equal to the sum of the 1000th powers of its digits).

Chapter 6

Complex Numbers

We all know that there are simple quadratic equations, such as $x^2 + 1 = 0$, that have no real solutions. In order to provide a notation with which to discuss such equations, we introduce a symbol i, and define

$$i^2 = -1.$$

A *complex number* is defined to be a symbol $a + bi$, where a, b are real numbers. If $z = a + bi$, we call a the *real part* of z and b the *imaginary part,* and write

$$a = Re(z), \quad b = Im(z).$$

We define addition and multiplication of complex numbers by the rules

$$\text{addition:} \quad (a+bi) + (c+di) = a+c+(b+d)i$$
$$\text{multiplication:} \quad (a+bi)(c+di) = ac - bd + (ad + bc)i.$$

Notice that in the multiplication rule, we multiply out the brackets in the usual way, and replace the i^2 by -1. For example, $(1+2i)(3-i) = 5 + 5i$ and $(a+bi)(a-bi) = a^2 + b^2$.

It is also possible to subtract complex numbers:

$$(a+bi) - (c+di) = a - c + (b-d)i$$

and, less obviously, to divide them: provided c, d are not both 0,

$$\frac{a+bi}{c+di} = \frac{(a+bi)(c-di)}{(c+di)(c-di)} = \frac{ac+bd}{c^2+d^2} + \left(\frac{bc-ad}{c^2+d^2}\right)i.$$

For example, $\frac{1-i}{1+i} = \frac{(1-i)(1-i)}{(1+i)(1-i)} = \frac{-2i}{2} = -i$.

We write $\mathbb{C}$ for the set of all complex numbers. Notice that if a and b are real numbers, then

$$(a+0i) + (b+0i) = a + b + 0i, \text{ and}$$
$$(a+0i)(b+0i) = ab + 0i,$$

so the complex numbers of the form $a + 0i$ add and multiply together just like the real numbers. If we identify the complex number $a + 0i$ with the real number a, we see that $\mathbb{R} \subseteq \mathbb{C}$.

Notice that every quadratic equation $ax^2 + bx + c = 0$ (where $a, b, c \in \mathbb{R}$) has roots in $\mathbb{C}$. For by the famous formula you will be familiar with, the roots are

$$\frac{1}{2a}\left(-b \pm \sqrt{b^2 - 4ac}\right).$$

If $b^2 \geq 4ac$ these roots lie in $\mathbb{R}$, while if $b^2 < 4ac$ they are the complex numbers $\frac{-b}{2a} \pm \frac{\sqrt{4ac - b^2}}{2a} i$.

It is straightforward to check from the definitions of addition and multiplication that the complex numbers obey the rules (1), (2) and (3) of Rules 2.1. The least obvious of these is the multiplcation rule in (2), that $(uv)w = u(vw)$ for all complex numbers u, v, w. Just to make sure you do this, I have set it as Exercise 1 at the end of the chapter.

Geometrical Representation of Complex Numbers

It turns out to be a very fruitful idea to represent complex numbers by points in the xy-plane. This is done in a natural way — the complex number $a + bi$ is represented by the point in the plane with coordinates (a, b). For example, i is represented by $(0, 1)$; $1 - i$ by $(1, -1)$; and so on:

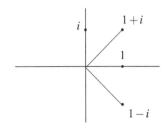

If $z = a + bi$, we define $\bar{z} = a - bi$ and call this the *complex conjugate* of z. Also, the *modulus* of $z = a + bi$ is the distance from the origin to the point (a, b) representing z. It is written as $|z|$. Thus,

$$|z| = \sqrt{a^2 + b^2}.$$

Notice that

$$z\bar{z} = (a + bi)(a - bi) = a^2 + b^2 = |z|^2.$$

The *argument* of z is the angle θ between the x-axis and the line joining 0 to z, measured in the counterclockwise direction:

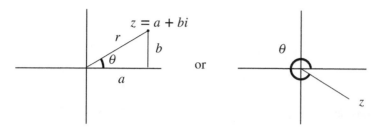

If $z = a + bi$ and $|z| = r$, then we see that $a = r\cos\theta, b = r\sin\theta$, so

$$z = r(\cos\theta + i\sin\theta).$$

This is known as the *polar form* of the complex number z.

Example 6.1

The polar forms of $i, -1, 1+i$ and $1-i$ are

$$i = 1\left(\cos\tfrac{\pi}{2} + i\sin\tfrac{\pi}{2}\right), \quad -1 = 1(\cos\pi + i\sin\pi),$$
$$1+i = \sqrt{2}\left(\cos\tfrac{\pi}{4} + i\sin\tfrac{\pi}{4}\right), \quad 1-i = \sqrt{2}\left(\cos\tfrac{7\pi}{4} + i\sin\tfrac{7\pi}{4}\right).$$

Let $z = r(\cos\theta + i\sin\theta)$. Notice that

$$\cos\theta + i\sin\theta = \cos(\theta + 2\pi) + i\sin(\theta + 2\pi)$$
$$= \cos(\theta + 4\pi) + i\sin(\theta + 4\pi) = \ldots,$$

so multiples of 2π can be added to θ (or subtracted from θ) without changing z. Thus, z has many different arguments. There is, however, a unique value of the argument of z in the range $-\pi < \theta \leq \pi$, and this is called the *principal argument* of z, written $arg(z)$. For example, $arg(1-i) = -\tfrac{\pi}{4}$.

De Moivre's Theorem

The xy-plane, representing the set of complex numbers as just described, is known as the *Argand diagram;* it is also sometimes called simply the *complex plane.*

The significance of the geometrical representation of complex numbers begins to become apparent in the next result, which shows that complex multiplication has a simple and natural geometric interpretation.

THEOREM 6.1 (De Moivre's Theorem)

Let z_1, z_2 be complex numbers with polar forms

$$z_1 = r_1\left(\cos\theta_1 + i\sin\theta_1\right), \quad z_2 = r_2\left(\cos\theta_2 + i\sin\theta_2\right).$$

Then the product

$$z_1 z_2 = r_1 r_2 \left(\cos(\theta_1 + \theta_2) + i \sin(\theta_1 + \theta_2) \right).$$

In other words, $z_1 z_2$ has modulus $r_1 r_2$ and argument $\theta_1 + \theta_2$.

PROOF   We have

$$\begin{aligned}
z_1 z_2 &= r_1 r_2 \left(\cos\theta_1 + i\sin\theta_1 \right) \left(\cos\theta_2 + i\sin\theta_2 \right) \\
&= r_1 r_2 \left(\cos\theta_1 \cos\theta_2 - \sin\theta_1 \sin\theta_2 + i \left(\cos\theta_1 \sin\theta_2 + \sin\theta_1 \cos\theta_2 \right) \right) \\
&= r_1 r_2 \left(\cos(\theta_1 + \theta_2) + i\sin(\theta_1 + \theta_2) \right). \quad \blacksquare
\end{aligned}$$

De Moivre's Theorem says that multiplying a complex number z by $\cos\theta + i\sin\theta$ rotates z counterclockwise through the angle θ; for example, multiplication by i rotates z through $\frac{\pi}{2}$:

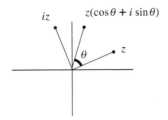

We now deduce a significant consequence of De Moivre's Theorem.

PROPOSITION 6.1
Let $z = r(\cos\theta + i\sin\theta)$, and let n be a positive integer. Then
(i) $z^n = r^n(\cos n\theta + i\sin n\theta)$, and
(ii) $z^{-n} = r^{-n}(\cos n\theta - i\sin n\theta)$.

PROOF   (i) Applying Theorem 6.1 with $z_1 = z_2 = z$ gives

$$z^2 = zz = rr(\cos(\theta + \theta) + i\sin(\theta + \theta)) = r^2(\cos 2\theta + i\sin 2\theta).$$

Repeating, we get

$$z^n = r\ldots r(\cos(\theta + \cdots + \theta) + i\sin(\theta + \cdots + \theta)) = r^n(\cos n\theta + i\sin n\theta).$$

(ii) First observe that

$$\begin{aligned}
z^{-1} = \frac{1}{z} &= \frac{1}{r(\cos\theta + i\sin\theta)} = \frac{1}{r} \frac{\cos\theta - i\sin\theta}{(\cos\theta + i\sin\theta)(\cos\theta - i\sin\theta)} \\
&= \frac{1}{r}(\cos\theta - i\sin\theta).
\end{aligned}$$

Hence $z^{-1} = r^{-1}(\cos(-\theta) + i\sin(-\theta))$, which proves the result for $n = 1$. And, for general n, we simply note that $z^{-n} = (z^{-1})^n$, which by part (i) is equal to $(r^{-1})^n(\cos(-n\theta) + i\sin(-n\theta))$, hence to $(r^{-n})(\cos(n\theta) - i\sin(n\theta))$. ∎

We now give a few examples illustrating the power of De Moivre's Theorem.

Example 6.2
Calculate $(-\sqrt{3} + i)^7$.

Answer We first find the polar form of $z = -\sqrt{3} + i$.

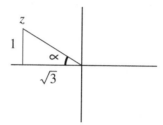

In the diagram, $\sin\alpha = \frac{1}{2}$, so $\alpha = \frac{\pi}{6}$. Hence $arg(z) = \frac{5\pi}{6}$. Also $|z| = 2$, so the polar form of z is

$$z = 2\left(\cos\frac{5\pi}{6} + i\sin\frac{5\pi}{6}\right).$$

Hence, by Proposition 6.1,

$$\left(-\sqrt{3} + i\right)^7 = 2^7\left(\cos\frac{35\pi}{6} + i\sin\frac{35\pi}{6}\right)$$

$$= 2^7\left(\cos\frac{-\pi}{6} + i\sin\frac{-\pi}{6}\right)$$

(subtracting 6π from the argument $\frac{35\pi}{6}$). Since $\cos\frac{-\pi}{6} = \frac{\sqrt{3}}{2}$ and $\sin\frac{-\pi}{6} = -\frac{1}{2}$, this gives

$$\left(-\sqrt{3} + i\right)^7 = 2^6\left(\sqrt{3} - i\right).$$

Example 6.3
Find a complex number w such that $w^2 = -\sqrt{3} + i$ (i.e., find a complex square root of $-\sqrt{3} + i$).

Answer From the previous solution, $-\sqrt{3} + i = 2(\cos\frac{5\pi}{6} + i\sin\frac{5\pi}{6})$. So if we define

$$w = \sqrt{2}\left(\cos\frac{5\pi}{12} + i\sin\frac{5\pi}{12}\right),$$

then by Proposition 6.1, $w^2 = -\sqrt{3}+i$. Note that $\sqrt{2}(\cos(\frac{5\pi}{12}+\pi)+i\sin(\frac{5\pi}{12}+\pi))$ works equally well; by Theorem 6.1 this is equal to $w(\cos\pi + i\sin\pi) = -w$.

Example 6.4

In this example we find a formula for $\cos 3\theta$ in terms of $\cos\theta$.

We begin with the equation

$$\cos 3\theta + i\sin 3\theta = (\cos\theta + i\sin\theta)^3.$$

Writing $c = \cos\theta, s = \sin\theta$, and expanding the cube, we get

$$\cos 3\theta + i\sin 3\theta = c^3 + 3c^2 si + 3cs^2 i^2 + s^3 i^3 = c^3 - 3cs^2 + i\left(3c^2 s - s^3\right).$$

Equating real parts, we have $\cos 3\theta = c^3 - 3cs^2$. Also $c^2 + s^2 = \cos^2\theta + \sin^2\theta = 1$, so $s^2 = 1 - c^2$, and therefore

$$\cos 3\theta = c^3 - 3c\left(1-c^2\right) = 4c^3 - 3c.$$

That is,

$$\cos 3\theta = 4\cos^3\theta - 3\cos\theta.$$

Example 6.5

We now use the previous example to find a cubic equation having $\cos\frac{\pi}{9}$ as a root.

Putting $\theta = \frac{\pi}{9}$ and $c = \cos\frac{\pi}{9}$, Example 6.4 gives

$$\cos 3\theta = 4c^3 - 3c.$$

However, $\cos 3\theta = \cos\frac{\pi}{3} = \frac{1}{2}$. Hence $\frac{1}{2} = 4c^3 - 3c$. In other words, $c = \cos\frac{\pi}{9}$ is a root of the cubic equation

$$8x^3 - 6x - 1 = 0.$$

Note that if $\phi = \frac{\pi}{9} + \frac{2\pi}{3}$ or $\frac{\pi}{9} + \frac{4\pi}{3}$, then $\cos 3\phi = \frac{1}{2}$, and hence the above argument shows $\cos\phi$ is also a root of this cubic equation. The roots of $8x^3 - 6x - 1 = 0$ are therefore $\cos\frac{\pi}{9}, \cos\frac{7\pi}{9}$ and $\cos\frac{13\pi}{9}$.

The $e^{i\theta}$ Notation

It is somewhat cumbersome to keep writing $\cos\theta + i\sin\theta$ in our notation for complex numbers. We therefore introduce a rather more compact notation by defining

$$e^{i\theta} = \cos\theta + i\sin\theta$$

for any real number θ. (This equation turns out to be very significant when $e^{i\theta}$ is regarded as an exponential function, but for now it is simply the definition of the symbol $e^{i\theta}$.)

For example,

$$e^{2\pi i} = 1, \quad e^{\pi i} = -1, \quad e^{\frac{\pi}{2}i} = i, \quad e^{\frac{\pi}{4}i} = \frac{1}{\sqrt{2}}(1+i).$$

Also, for any integer k,

$$e^{i\theta} = e^{i(\theta + 2k\pi)}.$$

Each of the complex numbers $e^{i\theta}$ has modulus 1, and the set consisting of all of them is the *unit circle* in the Argand diagram — that is, the circle of radius 1 centered at the origin:

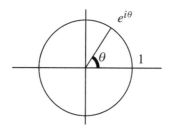

The polar form of a complex number z can now be written as

$$z = re^{i\theta}$$

where $r = |z|$ and $\theta = arg(z)$. For example,

$$-\sqrt{3} + i = 2e^{\frac{5\pi i}{6}}.$$

De Moivre's Theorem 6.1 implies that

$$e^{i\theta} e^{i\phi} = e^{i(\theta+\phi)},$$

and Proposition 6.1 says that for any integer n,

$$\left(e^{i\theta}\right)^n = e^{in\theta}.$$

From these facts we begin to see some of the significance emerging behind the definition of $e^{i\theta}$.

PROPOSITION 6.2

(i) If $z = re^{i\theta}$ then $\bar{z} = re^{-i\theta}$.
(ii) Let $z = re^{i\theta}, w = se^{i\phi}$ in polar form. Then $z = w$ if and only if both $r = s$ and $\theta - \phi = 2k\pi$ with $k \in \mathbb{Z}$.

PROOF   (i) We have $z = r(\cos\theta + i\sin\theta)$, so $\bar{z} = r(\cos\theta - i\sin\theta) = r(\cos(-\theta) + i\sin(-\theta)) = re^{-i\theta}$.

(ii) If $r = s$ and $\theta - \phi = 2k\pi$ with $k \in \mathbb{Z}$, then

$$z = re^{i\theta} = se^{i(\phi + 2k\pi)} = se^{i\phi} = w.$$

This does the "right to left" implication.

For the "left to right" implication, suppose $z = w$. Then $|z| = |w|$, so $r = s$ and also $e^{i\theta} = e^{i\phi}$. Now

$$e^{i\theta} = e^{i\phi} \Rightarrow e^{i\theta}e^{-i\phi} = e^{i\phi}e^{-i\phi} \Rightarrow e^{i(\theta - \phi)} = 1$$
$$\Rightarrow \cos(\theta - \phi) = 1, \ \sin(\theta - \phi) = 0 \Rightarrow \theta - \phi = 2k\pi \text{ with } k \in \mathbb{Z}. \quad \blacksquare$$

Roots of Unity

Consider the equation

$$z^3 = 1.$$

This is easy enough to solve: rewriting it as $z^3 - 1 = 0$, and factorizing this as $(z - 1)(z^2 + z + 1) = 0$, we see that the roots are

$$1, \ -\frac{1}{2} + \frac{\sqrt{3}}{2}i, \ -\frac{1}{2} - \frac{\sqrt{3}}{2}i.$$

These complex numbers have polar forms

$$1, \ e^{\frac{2\pi i}{3}}, \ e^{\frac{4\pi i}{3}}.$$

In other words, they are evenly spaced on the unit circle like this:

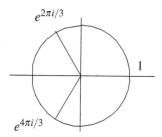

These three complex numbers are called the *cube roots of unity*.

More generally, if n is a positive integer, then the complex numbers that satisfy the equation

$$z^n = 1$$

are called the nth *roots of unity*.

PROPOSITION 6.3

Let n be a positive integer and define $w = e^{\frac{2\pi i}{n}}$. Then the nth roots of unity are the n complex numbers

$$1, w, w^2, \ldots, w^{n-1}$$

(i.e., $1, e^{\frac{2\pi i}{n}}, e^{\frac{4\pi i}{n}}, \ldots, e^{\frac{2(n-1)\pi i}{n}}$). They are evenly spaced around the unit circle.

PROOF Let $z = re^{i\theta}$ be an nth root of unity. Then

$$1 = z^n = r^n e^{ni\theta}.$$

From Proposition 6.2(ii) it follows that $r = 1$ and $n\theta = 2k\pi$ with $k \in \mathbb{Z}$. Therefore, $\theta = \frac{2k\pi}{n}$, and so $z = e^{\frac{2k\pi i}{n}} = w^k$.

Thus every n^{th} root of unity is a power of w. On the other hand, any power w^k is an nth root of unity, since

$$\left(w^k\right)^n = w^{nk} = \left(e^{\frac{2\pi i}{n}}\right)^{nk} = \left(e^{2\pi i}\right)^k = 1.$$

The complex numbers

$$1, w, w^2, \ldots, w^{n-1}$$

are all the distinct powers of w (since $w^n = 1, w^{n+1} = w$, etc.). Hence, these are the nth roots of unity. ∎

Example 6.6

The fourth roots of unity are $1, e^{\frac{i\pi}{2}}, e^{i\pi}, e^{\frac{3i\pi}{2}}$, which are just

$$1, i, -1, -i.$$

The sixth roots of unity are $1, e^{\frac{i\pi}{3}}, e^{\frac{2i\pi}{3}}, -1, e^{\frac{4i\pi}{3}}$ and $e^{\frac{5i\pi}{3}}$; these are the corners of a regular hexagon drawn inside the unit circle.

We can use the nth roots of unity to find the nth roots of any complex number. Here is an example.

Example 6.7

Find all solutions of the equation

$$z^5 = -\sqrt{3} + i.$$

(In other words, find all the fifth roots of $-\sqrt{3} + i$.)

Answer Let $p = -\sqrt{3} + i$. Recall that $p = 2e^{\frac{5\pi i}{6}}$. One of the fifth roots of this is clearly

$$\alpha = 2^{\frac{1}{5}} e^{\frac{\pi i}{6}}$$

(where of course $2^{\frac{1}{5}}$ is the real fifth root of 2). If w is a fifth root of unity, then $(\alpha w)^5 = \alpha^5 w^5 = \alpha^5 = z$, so αw is also a fifth root of p. Thus we have found the following 5 fifth roots of $-\sqrt{3} + i$:

$$\alpha, \alpha e^{\frac{2\pi i}{5}}, \alpha e^{\frac{4\pi i}{5}}, \alpha e^{\frac{6\pi i}{5}}, \alpha e^{\frac{8\pi i}{5}}.$$

These are in fact all the fifth roots of p: for if β is any fifth root of p, then $\beta^5 = \alpha^5 = p$, so $(\frac{\beta}{\alpha})^5 = 1$, which means that $\frac{\beta}{\alpha} = w$ is a fifth root of unity, and hence $\beta = \alpha w$ is in the above list.

We conclude that the fifth roots of $-\sqrt{3} + i$ are

$$2^{\frac{1}{5}} e^{\frac{\pi i}{6}}, 2^{\frac{1}{5}} e^{\frac{17\pi i}{30}}, 2^{\frac{1}{5}} e^{\frac{29\pi i}{30}}, 2^{\frac{1}{5}} e^{\frac{41\pi i}{30}}, 2^{\frac{1}{5}} e^{\frac{53\pi i}{30}}.$$

In general, the above method shows that if one of the nth roots of a complex number is β, then the others are $\beta w, \beta w^2, \ldots, \beta w^{n-1}$ where $w = e^{\frac{2\pi i}{n}}$.

Exercises for Chapter 6

1. Prove the following facts about complex numbers:

   (a) $u + v = v + u$ for all $u, v \in \mathbb{C}$.

   (b) $uv = vu$ for all $u, v \in \mathbb{C}$.

   (c) $(u + v) + w = (u + v) + w$ for all $u, v, w \in \mathbb{C}$.

   (d) $u(v + w) = uv + uw$ for all $u, v, w \in \mathbb{C}$.

   (e) $u(vw) = (uv)w$ for all $u, v, w \in \mathbb{C}$.

2. Prove the following, for all $u, v \in \mathbb{C}$:

   (a) $\overline{u + v} = \bar{u} + \bar{v}$.

   (b) $\overline{uv} = \bar{u}\bar{v}$.

   (c) $|u|^2 = u\bar{u}$.

   (d) $|uv| = |u||v|$.

3. (a) Find the real and imaginary parts of $(\sqrt{3} - i)^{10}$ and $(\sqrt{3} - i)^{-7}$. For which values of n is $(\sqrt{3} - i)^n$ real?

   (b) What is $\sqrt{i}$?

   (c) Find all the tenth roots of i. Which one is nearest to i in the Argand diagram?

   (d) Find the seven roots of the equation $z^7 - \sqrt{3} + i = 0$. Which one of these roots is closest to the imaginary axis?

4. Prove the "Triangle Inequality" for complex numbers: $|u + v| \le |u| + |v|$ for all $u, v \in \mathbb{C}$.

5. Let z be a non-zero complex number. Prove that the three cube roots of z are the corners of an equilateral triangle in the Argand diagram.

6. Express $\frac{1+i}{\sqrt{3}+i}$ in the form $x + iy$, where $x, y \in \mathbb{R}$. By writing each of $1 + i$ and $\sqrt{3} + i$ in polar form, deduce that

$$\cos \frac{\pi}{12} = \frac{\sqrt{3}+1}{2\sqrt{2}}, \quad \sin \frac{\pi}{12} = \frac{\sqrt{3}-1}{2\sqrt{2}}.$$

7. (a) Show that $x^5 - 1 = (x - 1)(x^4 + x^3 + x^2 + x + 1)$. Deduce that if $\omega = e^{2\pi i/5}$ then $\omega^4 + \omega^3 + \omega^2 + \omega + 1 = 0$.

   (b) Let $\alpha = 2\cos \frac{2\pi}{5}$ and $\beta = 2\cos \frac{4\pi}{5}$. Show that $\alpha = \omega + \omega^4$ and $\beta = \omega^2 + \omega^3$. Find a quadratic equation with roots α, β. Hence show that

$$\cos \frac{2\pi}{5} = \frac{1}{4}\left(\sqrt{5} - 1\right).$$

8. Find a formula for $\cos 4\theta$ in terms of $\cos \theta$. Hence write down a quartic equation (i.e., an equation of degree 4) that has $\cos \frac{\pi}{12}$ as a root. What are the other roots of your equation?

9. Find all complex numbers z such that $|z| = |\sqrt{2} + z| = 1$. Prove that each of these satisfies $z^8 = 1$.

10. Prove that there is no complex number z such that $|z| = |z + i\sqrt{5}| = 1$.

11. Show that if w is an nth root of unity, then $\bar{w} = \frac{1}{w}$. Deduce that

$$\overline{(1-w)}^n = (w-1)^n.$$

Hence show that $(1-w)^{2n}$ is real.

12. Let n be a positive integer, and let $z \in \mathbb{C}$ satisfy the equation

$$(z-1)^n + (z+1)^n = 0.$$

(a) Show that $z = \frac{1+w}{1-w}$ for some $w \in \mathbb{C}$ such that $w^n = -1$.

(b) Show that $w\bar{w} = 1$.

(c) Deduce that z lies on the imaginary axis.

13. Critic Ivor Smallbrain is discussing the film *Sets, Lines and Videotape* with his two chief editors, Sir Giles Tantrum and Lord Overthetop. They are sitting at a circular table of radius 1. Ivor is bored and notices in a daydream that he can draw real and imaginary axes, with origin at the center of the table, in such a way that Tantrum is represented by a certain complex number z and Overthetop is represented by the complex number $z + 1$. Breaking out of his daydream, Ivor suddenly exclaims, "You are both sixth roots of 1!"

Prove that Ivor is correct, despite the incredulous editorial glares.

Chapter 7

Polynomial Equations

Expressions like $x^2 - 3x$ or $-7x^{102} + (3-i)x^{17} - 7$, or more generally,

$$p(x) = a_n x^n + a_{n-1} x^{n-1} + \cdots + a_1 x + a_0,$$

where the coefficients $a_0, a_1, \ldots, a_n$ are complex numbers, are called *polynomials* in x. A *polynomial equation* is an equation of the form

$$p(x) = 0$$

where $p(x)$ is a polynomial. The *degree* of such an equation is the highest power of x that appears with a non-zero coefficient.

For example, equations of degree 1 take the form $ax + b = 0$ and are also known as *linear* equations; degree 2 equations $ax^2 + bx + c = 0$ are *quadratic* equations; degree 3 equations are *cubic* equations; degree 4 are *quartic* equations; degree 5 are *quintic* equations; and so on.

A complex number α is said to be a *root* of the polynomial equation $p(x) = 0$ if $p(\alpha) = 0$: in other words, when α is substituted for x, $p(x)$ becomes equal to 0. For example, 1 is a root of the cubic equation $x^3 - 3x + 2 = 0$.

The search for formulae for the roots of polynomial equations was one of the driving forces in mathematics from the time of the Greeks until the nineteenth century. Let us now taste a tiny flavour of this huge subject, in the hope that appetites are whetted for more.

It is obvious that any linear equation $ax + b = 0$ has exactly one root, namely $-\frac{b}{a}$. We are also familiar with the fact that any quadratic equation $ax^2 + bx + c = 0$ has roots in $\mathbb{C}$, given by the formula $\frac{1}{2a}(-b \pm \sqrt{b^2 - 4ac})$.

Things are less clear for cubic equations. Indeed, while the formula for the roots of a quadratic was known to the Greeks, it was not until the sixteenth century that a method for finding the roots of a cubic was found by the Italian mathematicians Scipio Ferreo, Tartaglia and Cardan. Here is their method.

Solution of Cubic Equations

Consider the cubic equation

$$x^3 + ax^2 + bx + c = 0. \tag{7.1}$$

The first step is to get rid of the x^2 term. This is easily done: put $y = x + \frac{a}{3}$. Then $y^3 = (x + \frac{a}{3})^3 = x^3 + ax^2 + \frac{a^2}{3}x + \frac{a^3}{27}$, so Equation (7.1) becomes $y^3 + b'y + c' = 0$ for some b', c'. Write this equation as

$$y^3 + 3hy + k = 0. \tag{7.2}$$

(The coefficients $3h$ and k can easily be worked out, given a, b, c.)

Here comes the clever part. Write $y = u + v$. Then

$$y^3 = (u+v)^3 = u^3 + v^3 + 3u^2v + 3uv^2 = u^3 + v^3 + 3uv(u+v)$$
$$= u^3 + v^3 + 3uvy.$$

Hence, the cubic equation

$$y^3 - 3uvy - (u^3 + v^3) = 0. \tag{7.3}$$

has $u + v$ as a root.

Our aim now is to find u and v so that the coefficients in Equations (7.2) and (7.3) are matched up. To match the coefficients, we require

$$h = -uv, \quad k = -\left(u^3 + v^3\right). \tag{7.4}$$

From the first of these equations we have $v^3 = \frac{-h^3}{u^3}$, hence the second equation gives $u^3 - \frac{h^3}{u^3} = -k$, so

$$u^6 + ku^3 - h^3 = 0. \tag{7.5}$$

This is just a quadratic equation for u^3, and a solution is

$$u^3 = \frac{1}{2}\left(-k + \sqrt{k^2 + 4h^3}\right).$$

Then from (7.4),

$$v^3 = -k - u^3 = \frac{1}{2}\left(-k - \sqrt{k^2 + 4h^3}\right).$$

As $y = u + v$, we have obtained the following formula for the roots of the cubic (7.2):

$$\sqrt[3]{\frac{1}{2}\left(-k + \sqrt{k^2 + 4h^3}\right)} + \sqrt[3]{\frac{1}{2}\left(-k - \sqrt{k^2 + 4h^3}\right)}.$$

Since a complex number has three cube roots, and there are two cube roots to be chosen, it seems that there are nine possible values for this formula. However, the equation $uv = -h$ implies that $v = \frac{-h}{u}$, and hence there are only three roots of (7.2), these being $u - \frac{h}{u}$ for each of the three choices for u.

Specifically, if u is one of the cube roots of $\frac{1}{2}(-k + \sqrt{k^2 + 4h^3})$, the other cube roots are $u\omega, u\omega^2$ where $\omega = e^{\frac{2\pi i}{3}}$, and so the roots of the cubic equation (7.2) are

$$u - \frac{h}{u}, \quad u\omega - \frac{h\omega^2}{u}, \quad u\omega^2 - \frac{h\omega}{u}.$$

Once we know the roots of (7.2), we can of course write down the roots of the general cubic (7.1), since $x = y - \frac{a}{3}$.

Let us illustrate this method with a couple of examples.

Example 7.1

(1) Consider the cubic equation $x^3 - 6x - 9 = 0$. This is (7.2) with $h = -2, k = -9$, so $\frac{1}{2}(-k + \sqrt{k^2 + 4h^3}) = \frac{1}{2}(9 + \sqrt{49}) = 8$. Hence, taking $u = 2$, we see that the roots are $3, 2\omega + \omega^2, 2\omega^2 + \omega$. As $\omega = \frac{1}{2}(-1 + i\sqrt{3})$ and $\omega^2 = \frac{1}{2}(-1 - i\sqrt{3})$, these roots are

$$3, \quad \frac{1}{2}\left(-3 + i\sqrt{3}\right), \quad \frac{1}{2}\left(-3 - i\sqrt{3}\right).$$

(Of course these could easily have been worked out by cleverly spotting that 3 is a root and factorizing the equation as $(x - 3)(x^2 + 3x + 3) = 0$.)

(2) Consider the equation $x^3 - 6x - 40 = 0$. The above formula gives roots of the form

$$\sqrt[3]{20 + 14\sqrt{2}} + \sqrt[3]{20 - 14\sqrt{2}}.$$

However, we cleverly also spot that 4 is a root. What is going on?

In fact, nothing very mysterious is going on. The real cube root of $20 \pm 14\sqrt{2}$ is $2 \pm \sqrt{2}$, as can be seen by cubing the latter. Hence, the roots of the cubic are

$$\left(2 + \sqrt{2}\right) + \left(2 - \sqrt{2}\right) = 4,$$
$$\left(2 + \sqrt{2}\right)\omega + \left(2 - \sqrt{2}\right)\omega^2 = -2 + i\sqrt{6},$$
$$\left(2 + \sqrt{2}\right)\omega^2 + \left(2 - \sqrt{2}\right)\omega = -2 - i\sqrt{6}.$$

Higher Degrees

Not long after the solution of the cubic, Ferrari, a pupil of Cardan, showed how to obtain a formula for the roots of a general quartic (degree 4) equation. The next step, naturally enough, was the quintic. However, several hundred years passed without anyone finding a formula for the roots of a general quintic equation.

There was a good reason for this. There is no such formula. Nor is there a formula for equations of degree greater than 5. This amazing fact was first established in the early 19th century by the Danish mathematician Abel (who died at age 26), after which the Frenchman Galois (who died at age 21) built an entirely new theory of equations, linking them to the then-recent subject of group theory, which not only explained the non-existence of formulae, but laid the foundations of a whole edifice of algebra and number theory known as *Galois theory,* a major area of modern-day research. If you get a chance, take a course in Galois theory during the rest of your studies in mathematics — you won't regret it!

The Fundamental Theorem of Algebra

So, there is no formula for the roots of a polynomial equation of degree 5 or more. We are therefore led to the troubling question: can we be sure that such an equation actually has a root in the complex numbers?

The answer to this is yes, we can be sure. This is a famous theorem of another great mathematician — perhaps the greatest of all — Gauss:

THEOREM 7.1 Fundamental Theorem of Algebra
Every polynomial equation of degree at least 1 has a root in $\mathbb{C}$.

This is really a rather amazing result. After all, we introduced complex numbers just to be able to talk about roots of quadratics like $x^2 + 1 = 0$, and we find ourselves with a system that contains roots of *all* polynomial equations.

There are many different proofs of the Fundamental Theorem of Algebra available — Gauss himself found five during his lifetime. Probably the proofs that are easiest to understand are those using various basic results in the subject of complex analysis, and most undergraduate courses on this topic would include a proof of the Fundamental Theorem of Algebra. In Chapter 24 we shall give a proof of the theorem for real polynomials of odd degree.

The Fundamental Theorem of Algebra has many consequences. We shall give just a couple here. Let $p(x)$ be a polynomial of degree n. By the Fundamental Theorem 7.1, $p(x)$ has a root in $\mathbb{C}$, say α_1. It is rather easy to see from this (but we won't prove it here) that there is another polynomial $q(x)$, of degree $n-1$, such that

$$p(x) = (x-\alpha_1)q(x).$$

By Theorem 7.1, $q(x)$ also has a root in $\mathbb{C}$, say α_2, so as above there is a polynomial $r(x)$ of degree $n-2$ such that

$$p(x) = (x-\alpha_1)(x-\alpha_2)r(x).$$

Repeat this argument until we get down to a polynomial of degree 1. We thus obtain a factorization

$$p(x) = a(x-\alpha_1)(x-\alpha_2)\ldots(x-\alpha_n),$$

where a is the coefficient of x^n, and $\alpha_1,\ldots,\alpha_n$ are the roots of $p(x)$. These may of course not all be different, but there are precisely n of them if we count repeats.

Summarizing:

THEOREM 7.2

Every polynomial of degree n factorizes as a product of linear polynomials and has exactly n roots in $\mathbb{C}$ (counting repeats).

Next, we briefly consider *real* polynomial equations — that is, equations of the form

$$p(x) = a_n x^n + a_{n-1}x^{n-1} + \cdots + a_1 x + a_0 = 0,$$

where all the coefficients $a_0, a_1, \ldots, a_n$ are real numbers.

Of course not all the roots of such an equation need be real (think of $x^2+1 = 0$). However, we can say something interesting about the roots.

Let $\alpha \in \mathbb{C}$ be a root of the real polynomial equation $p(x) = 0$. Thus,

$$p(\alpha) = a_n \alpha^n + a_{n-1}\alpha^{n-1} + \cdots + a_1 \alpha + a_0 = 0.$$

Consider the complex conjugate $\bar{\alpha}$. We shall show this is also a root. To see this, observe first that $\overline{\alpha^2} = \bar{\alpha}^2$ (just apply Exercise 1(b) of Chapter 6 with $u = v = \alpha$); likewise, $\overline{\alpha^3} = \bar{\alpha}^3, \ldots, \overline{\alpha^n} = \bar{\alpha}^n$. Also $\bar{a}_i = a_i$ for all i, since the a_i are all real. Consequently,

$$\begin{aligned}
p(\bar{\alpha}) &= a_n \bar{\alpha}^n + a_{n-1}\bar{\alpha}^{n-1} + \cdots + a_1 \bar{\alpha} + a_0 \\
&= a_n \overline{\alpha^n} + a_{n-1}\overline{\alpha^{n-1}} + \cdots + a_1 \bar{\alpha} + a_0 \\
&= \overline{a_n \alpha^n} + \overline{a_{n-1}\alpha^{n-1}} + \cdots + \overline{a_1 \alpha} + \overline{a_0} \\
&= \overline{p(\alpha)} = 0,
\end{aligned}$$

showing that $\bar{\alpha}$ is indeed a root.

Thus for a real polynomial equation $p(x) = 0$, the non-real roots appear in complex conjugate pairs $\alpha, \bar{\alpha}$. Say the real roots are $\beta_1, \ldots, \beta_k$ and the non-real roots are $\alpha_1, \bar{\alpha}_1, \ldots, \alpha_l, \bar{\alpha}_l$ (where $k + 2l = n$). Then, as discussed above,

$$p(x) = (x - \beta_1) \ldots (x - \beta_k)(x - \alpha_1)(x - \bar{\alpha}_1) \ldots (x - \alpha_l)(x - \bar{\alpha}_l).$$

Notice that $(x - \alpha_i)(x - \bar{\alpha}_i) = x^2 - (\alpha_i + \bar{\alpha}_i)x + \alpha_i \bar{\alpha}_i$, which is a quadratic with real coefficients. Thus, $p(x)$ factorizes as a product of real linear and real quadratic polynomials.

Summarizing:

THEOREM 7.3

Every real polynomial factorizes as a product of real linear and real quadratic polynomials and has its non-real roots appearing in complex conjugate pairs.

Relationships between Roots

Despite the fact that there is no general formula for the roots of a polynomial equation of degree 5 or more, there are still some interesting and useful relationships between the roots and the coefficients.

First consider a quadratic equation $x^2 + ax + b = 0$. If α, β are the roots, then $x^2 + ax + b = (x - \alpha)(x - \beta) = x^2 - (\alpha + \beta)x + \alpha\beta$, and hence, equating coefficients, we have

$$\alpha + \beta = -a, \quad \alpha\beta = b.$$

Likewise, for a cubic equation $x^3 + ax^2 + bx + c = 0$ with roots α, β, γ, we have $x^3 + ax^2 + bx + c = (x - \alpha)(x - \beta)(x - \gamma)$, hence

$$\alpha + \beta + \gamma = -a, \quad \alpha\beta + \alpha\gamma + \beta\gamma = b, \quad \alpha\beta\gamma = -c.$$

Applying this argument to an equation of degree n, we have

PROPOSITION 7.1

Let the roots of the equation

$$x^n + a_{n-1}x^{n-1} + \cdots + a_1x + a_0 = 0$$

be $\alpha_1, \alpha_2, \ldots, \alpha_n$. If s_1 denotes the sum of the roots, s_2 denotes the sum of all products of pairs of roots, s_3 denotes the sum of all products of *triples of roots, and so on, then*

$$s_1 = \alpha_1 + \cdots + \alpha_n = -a_{n-1},$$
$$s_2 = a_{n-2},$$
$$s_3 = -a_{n-3},$$
$$\cdots \quad \cdots$$
$$s_n = \alpha_1 \alpha_2 \ldots \alpha_n = (-1)^n a_0.$$

PROOF We have

$$x^n + a_{n-1}x^{n-1} + \cdots + a_1 x + a_0 = (x - \alpha_1)(x - \alpha_2)\ldots(x - \alpha_n).$$

If we multiply out the right-hand side, the coefficient of x^{n-1} is $-(\alpha_1 + \cdots + \alpha_n) = -s_1$, the coefficient of x^{n-2} is s_2, and so on. The result follows.
∎

Here are some examples of what can be done with this result.

Example 7.2

(1) Write down a cubic equation with roots $1+i, 1-i, 2$.

Answer If we call these three roots α_1, α_2, α_3, then $s_1 = \alpha_1 + \alpha_2 + \alpha_3 = 4$, $s_2 = \alpha_1\alpha_2 + \alpha_1\alpha_3 + \alpha_2\alpha_3 = (1+i)(1-i) + 2(1+i) + 2(1-i) = 6$ and $s_3 = \alpha_1\alpha_2\alpha_3 = 4$. Hence, a cubic with these roots is

$$x^3 - 4x^2 + 6x - 4 = 0.$$

(2) If α, β are the roots of the equation $x^2 - 5x + 9 = 0$, find a quadratic equation with roots α^2, β^2.

Answer Of course this could be done by using the formula to write down α and β, then squaring them, and so on; this would be rather tedious, and a much more elegant solution is as follows. From the above, we have

$$\alpha + \beta = 5, \quad \alpha\beta = 9.$$

Therefore, $\alpha^2 + \beta^2 = (\alpha + \beta)^2 - 2\alpha\beta = 5^2 - 18 = 7$ and $\alpha^2\beta^2 = 9^2 = 81$. We therefore know the sum and product of α^2 and β^2, so a quadratic having these as roots is $x^2 - 7x + 81 = 0$.

(3) Find the value of k if the roots of the cubic equation $x^3 + x^2 + 2x + k = 0$ are in geometric progression.

Answer Saying that the roots are in geometric progression means that they are of the form $\alpha, \alpha r, \alpha r^2$ for some r. We then have

$$\alpha\left(1 + r + r^2\right) = -1, \quad \alpha^2\left(r + r^2 + r^3\right) = 2, \quad \alpha^3 r^3 = -k.$$

Dividing the first two of these gives $\alpha r = -2$. Hence, the third gives $k = 8$.

Exercises for Chapter 7

1. (a) Find the (complex) roots of the quadratic equation $x^2 - 5x + 7 - i = 0$.

   (b) Find the roots of the quartic equation $x^4 + x^2 + 1 = 0$.

   (c) Find the roots of the equation $2x^4 - 4x^3 + 3x^2 + 2x - 2 = 0$, given that one of them is $1 + i$.

2. Use the method given in this chapter for solving cubics to find the roots of the equation $x^3 - 6x^2 + 13x - 12 = 0$.

   Now notice that 3 is one of the roots. Reconcile this with the roots you have found.

3. Solve $x^3 - 15x - 4 = 0$ using the method for solving cubics.

   Now cleverly spot an integer root. Deduce that

$$\cos\left(\frac{1}{3}\tan^{-1}\left(\frac{11}{2}\right)\right) = \frac{2}{\sqrt{5}}.$$

4. Factorize $x^5 + 1$ as a product of real linear and quadratic polynomials.

5. Show that $\cos\frac{2\pi}{9}$ is a root of the cubic equation $8x^3 - 6x + 1 = 0$.

   Find the other two roots and deduce that

$$\cos\frac{2\pi}{9} + \cos\frac{4\pi}{9} + \cos\frac{8\pi}{9} = 0$$

   and

$$\cos\frac{2\pi}{9} \cdot \cos\frac{4\pi}{9} \cdot \cos\frac{8\pi}{9} = -\frac{1}{8}.$$

6. (a) Factorize $x^{2n+1} - 1$ as a product of real linear and quadratic polynomials.

   (b) Write $x^{2n} + x^{2n-1} + \cdots + x + 1$ as a product of real quadratic polynomials.

   (c) Let $\omega = e^{2\pi i/2n+1}$. Show that $\sum \omega^{i+j} = 0$, where the sum is over all i and j from 1 to $2n+1$ such that $i < j$.

7. (a) Find the value of k such that the roots of $x^3 + 6x^2 + kx - 10 = 0$ are in arithmetical progression (i.e., are $\alpha, \alpha + d, \alpha + 2d$ for some d). Solve the equation for this value of k.

(b) If the roots of the equation $x^3 - x - 1 = 0$ are α, β, γ, find a cubic equation having roots $\alpha^2, \beta^2, \gamma^2$ and also a cubic equation having roots $\frac{1}{\alpha}, \frac{1}{\beta}, \frac{1}{\gamma}$.

(c) Given that the sum of two of the roots of the equation $x^3 + px^2 + p^2x + r = 0$ is 1, prove that $r = (p+1)(p^2 + p + 1)$.

(d) Solve the equation $x^4 - 3x^3 - 5x^2 + 17x - 6 = 0$, given that the sum of two of the roots is 5.

8. Critic Ivor Smallbrain and his friend Polly Gnomialle have entered a competition to try to qualify to join the UK team for the annual Film Critics Mathematical Olympiad. The qualification competition consists of one question. Here it is:

Find all real or complex solutions of the simultaneous equations

$$x + y + z = 3$$
$$x^2 + y^2 + z^2 = 3$$
$$x^3 + y^3 + z^3 = 3.$$

The first pair to correctly answer the question gets into the team. Ivor overhears his arch-rival Greta Picture, who has also entered the competition, whispering to her partner, "Let's look for a cubic equation which has roots x, y and z."

Can you help Ivor and Polly win?

Chapter 8

Induction

Consider the following three statements, each involving a general positive integer n:

(1) The sum of the first n odd numbers is equal to n^2.

(2) If $p > -1$ then $(1+p)^n \geq 1 + np$.

(3) The sum of the internal angles in an n-sided polygon is $(n-2)\pi$.

[A *polygon* is a closed figure with straight edges, such as a triangle (3 sides), a quadrilateral (4 sides), a pentagon (5 sides), etc.]

We can check that these statements are true for various specific values of n. For instance, (1) is true for $n = 2$ as $1 + 3 = 4 = 2^2$, and for $n = 3$ as $1 + 3 + 5 = 9 = 3^2$; statement (2) is true for $n = 1$ as $1 + p \geq 1 + p$, and for $n = 2$ as $(1+p)^2 = 1 + 2p + p^2 \geq 1 + 2p$; and (3) is true for $n = 3$ as the sum of the angles in a triangle is π, and for $n = 4$ as the sum of the angles in a quadrilateral is 2π.

But how do we go about trying to prove the truth of these statements for *all* values of n?

The answer is that we use the following basic principle. In it we denote by $P(n)$ a statement involving a positive integer n; for example, $P(n)$ could be any of statements (1), (2) or (3) above.

Principle of Mathematical Induction
Suppose that for each positive integer n we have a statement $P(n)$. If we prove the following two things:
   (a) $P(1)$ *is true;*
   (b) *for all n, if $P(n)$ is true then $P(n+1)$ is also true;*
then $P(n)$ is true for all positive integers n.

The logic behind this principle is clear: by (a), the first statement $P(1)$ is true. By (b) with $n = 1$, we know that $P(1) \Rightarrow P(2)$, hence $P(2)$ is true. By (b) with $n = 2$, $P(2) \Rightarrow P(3)$, hence $P(3)$ is true; and so on.

The principle may look a little strange at first sight, but a few examples should clarify matters.

Example 8.1

Let us try to prove statement (1) above using the Principle of Mathematical Induction. Here $P(n)$ is the statement that the sum of the first n odd numbers is n^2. In other words:

$$P(n) : 1 + 3 + 5 + \cdots + 2n - 1 = n^2.$$

We need to carry out parts (a) and (b) of the principle.

(a) $P(1)$ is true, since $1 = 1^2$.

(b) Suppose $P(n)$ is true. Then

$$1 + 3 + 5 + \cdots + 2n - 1 = n^2.$$

Adding $2n + 1$ to both sides gives

$$1 + 3 + 5 + \cdots + 2n - 1 + 2n + 1 = n^2 + 2n + 1 = (n+1)^2,$$

which is statement $P(n+1)$. Thus, we have shown that $P(n) \Rightarrow P(n+1)$.

We have now established parts (a) and (b). Hence by the Principle of Mathematical Induction, $P(n)$ is true for all positive integers n.

The phrase "Principle of Mathematical Induction" is quite a mouthful, and we usually use just the single word "induction" instead.

Example 8.2

Now let us prove statement (2) above by induction. Here, for n a positive integer $P(n)$ is the statement

$$P(n) : \text{ if } p > -1 \text{ then } (1+p)^n \geq 1 + np.$$

For (a), observe $P(1)$ is true, as $1 + p \geq 1 + p$.

For (b), suppose $P(n)$ is true, so $(1+p)^n \geq 1 + np$. Since $p > -1$, we know that $1 + p > 0$, so we can multiply both sides of the inequality by $1 + p$ (see Example 5.3) to obtain

$$(1+p)^{n+1} \geq (1+np)(1+p) = 1 + (n+1)p + np^2.$$

Since $np^2 \geq 0$, this implies that $(1+p)^{n+1} \geq 1 + (n+1)p$, which is statement $P(n+1)$. Thus we have shown $P(n) \Rightarrow P(n+1)$.

Therefore, by induction, $P(n)$ is true for all positive integers n.

Next we attempt to prove the statement (3) concerning n-sided polygons. There is a slight problem here. If we naturally enough let $P(n)$ be statement (3), then $P(n)$ makes sense only if $n \geq 3$; $P(1)$ and $P(2)$ make no sense, as there is no such thing as a 1-sided or 2-sided polygon. To take care of such a situation, we need a slightly modified Principle of Mathematical Induction:

Principle of Mathematical Induction II

Let k be an integer. Suppose that for each integer $n \geq k$ we have a statement $P(n)$. If we prove the following two things:
  (a) *$P(k)$ is true;*
  (b) *for all $n \geq k$, if $P(n)$ is true then $P(n+1)$ is also true;*
then $P(n)$ is true for all integers $n \geq k$.

The logic behind this is the same as explained before.

Example 8.3

Now we prove statement (3). Here we have $k = 3$ in the above principle, and for $n \geq 3$, $P(n)$ is the statement

$P(n)$: the sum of the internal angles in an n-sided polygon is $(n-2)\pi$.

For (a), observe that $P(3)$ is true, since the sum of the angles in a triangle is $\pi = (3-2)\pi$.

Now for (b). Suppose $P(n)$ is true. Consider an $(n+1)$-sided polygon with corners $A_1, A_2, \ldots, A_{n+1}$:

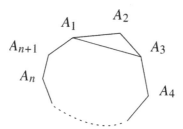

Draw the line A_1A_3. Then $A_1A_3A_4 \ldots A_{n+1}$ is an n-sided polygon. Since we are assuming $P(n)$ is true, the internal angles in this n-sided polygon add up to $(n-2)\pi$. From the picture we see that the sum of the angles in the $(n+1)$-sided polygon $A_1A_2 \ldots A_{n+1}$ is equal to the sum of those in $A_1A_3A_4 \ldots A_{n+1}$ plus the sum of those in the triangle $A_1A_2A_3$, hence is

$$(n-2)\pi + \pi = ((n+1)-2)\pi.$$

We have now shown that $P(n) \Rightarrow P(n+1)$. Hence, by induction, $P(n)$ is true for all $n \geq 3$.

The next example also uses the slightly modified Principle of Mathematical Induction II. In it, for a positive integer n we define

$$n! = n(n-1)(n-2)\ldots 3 \cdot 2 \cdot 1,$$

the product of all the integers between 1 and n. The symbol $n!$ is usually referred to as n *factorial*. By convention, we also define $0! = 1$.

Example 8.4

For which positive integers n is $2^n < n!$?

Answer Let $P(n)$ be the statement that $2^n < n!$. Observe that

$$2^1 > 1!, \; 2^2 > 2!, \; 2^3 > 3!, \; 2^4 < 4!, \; 2^5 < 5!,$$

so $P(1), P(2), P(3)$ are false, while $P(4), P(5)$ are true. Therefore, it seems sensible to try to prove $P(n)$ is true for all $n \geq 4$.

First, $P(4)$ is true, as observed above.

Now suppose n is an integer with $n \geq 4$, and $P(n)$ is true. Thus

$$2^n < n!$$

Multiplying both sides by 2, we get

$$2^{n+1} < 2(n!).$$

Since $2 < n+1$, we have $2(n!) < (n+1)n! = (n+1)!$ and hence $2^{n+1} < (n+1)!$. This shows that $P(n) \Rightarrow P(n+1)$. Therefore, by induction, $P(n)$ is true for all $n \geq 4$.

Guessing the Answer

Some problems cannot immediately be tackled using induction, but first require some intelligent guesswork. Here is an example.

Example 8.5

Find a formula for the sum

$$\frac{1}{1 \cdot 2} + \frac{1}{2 \cdot 3} + \cdots + \frac{1}{n(n+1)}.$$

Answer Calculate this sum for the first few values of n:

$$n = 1 : \frac{1}{1 \cdot 2} = \frac{1}{2},$$

$$n = 2 : \frac{1}{1 \cdot 2} + \frac{1}{2 \cdot 3} = \frac{1}{2} + \frac{1}{6} = \frac{2}{3},$$

$$n = 3 : \frac{1}{1 \cdot 2} + \frac{1}{2 \cdot 3} + \frac{1}{3 \cdot 4} = \frac{3}{4}.$$

We intelligently spot a pattern in these answers and guess that the sum of n terms is probably $\frac{n}{n+1}$. Hence we let $P(n)$ be the statement

$$P(n) : \frac{1}{1 \cdot 2} + \frac{1}{2 \cdot 3} + \cdots + \frac{1}{n(n+1)} = \frac{n}{n+1}$$

and attempt to prove $P(n)$ true for all $n \geq 1$ by induction.

First, $P(1)$ is true, as noted above.

Now assume $P(n)$ is true, so

$$\frac{1}{1 \cdot 2} + \frac{1}{2 \cdot 3} + \cdots + \frac{1}{n(n+1)} = \frac{n}{n+1}.$$

Adding $\frac{1}{(n+1)(n+2)}$ to both sides gives

$$\frac{1}{1 \cdot 2} + \cdots + \frac{1}{n(n+1)} + \frac{1}{(n+1)(n+2)} = \frac{n}{n+1} + \frac{1}{(n+1)(n+2)}$$

$$= \frac{n(n+2)+1}{(n+1)(n+2)} = \frac{n^2+2n+1}{(n+1)(n+2)} = \frac{(n+1)^2}{(n+1)(n+2)} = \frac{n+1}{n+2}.$$

Hence $P(n) \Rightarrow P(n+1)$. So, by induction, $P(n)$ is true for all $n \geq 1$.

The Σ Notation

Before proceeding with the next example, we introduce an important notation for writing down sums of many terms. If $f_1, f_2, \ldots, f_n$ are numbers, we abbreviate the sum of all of them by

$$f_1 + f_2 + \cdots + f_n = \sum_{r=1}^{n} f_r.$$

(The symbol Σ is the Greek capital letter "sigma," so this is often called the "sigma notation.") For example, setting $f_r = \frac{1}{r(r+1)}$, we have

$$\frac{1}{1 \cdot 2} + \frac{1}{2 \cdot 3} + \cdots + \frac{1}{n(n+1)} = \sum_{r=1}^{n} \frac{1}{r(r+1)}.$$

Thus, Example 8.5 says that

$$\sum_{r=1}^{n} \frac{1}{r(r+1)} = \frac{n}{n+1},$$

and Example 8.1 says

$$\sum_{r=1}^{n} (2r-1) = n^2.$$

Notice that if a, b, c are constants, then

$$\sum_{r=1}^{n} (af_r + bg_r + c) = a \sum_{r=1}^{n} f_r + b \sum_{r=1}^{n} g_r + cn, \tag{8.1}$$

since the left-hand side is equal to

$$(af_1 + bg_1 + c) + \cdots + (af_n + bg_n + c)$$
$$= a(f_1 + \cdots + f_n) + b(g_1 + \cdots + g_n) + (c + \cdots + c),$$

which is the right-hand side.

The equation (8.1) is quite useful for manipulating sums. Here is an elementary example using it.

Example 8.6

Find a formula for $\sum_{r=1}^{n} r \; (= 1 + 2 + \cdots + n)$.

Answer Write $s_n = \sum_{r=1}^{n} r$. By Example 8.1, $\sum_{r=1}^{n} (2r-1) = n^2$, so using (8.1),

$$n^2 = \sum_{r=1}^{n} (2r-1) = 2 \sum_{r=1}^{n} r - n = 2s_n - n.$$

Hence, $s_n = \frac{1}{2} n(n+1)$.

So we know the sum of the first n positive integers. What about the sum of the first n squares?

Example 8.7

Find a formula for $\sum_{r=1}^{n} r^2 \; (= 1^2 + 2^2 + \cdots + n^2)$.

Answer We first try to guess the answer (intelligently). The first few values $n = 1, 2, 3, 4$ give sums $1, 5, 14, 30$. It is not easy to guess a formula from these values, so yet a smidgeon more intelligence is required. The sum we are trying to find is the sum of n terms of a quadratic nature, so it seems reasonable to look for a formula for the sum which is a cubic in n, say $an^3 + bn^2 + cn + d$.

What should a, b, c, d be? Well, they have to fit in with the values of the sum for $n = 1, 2, 3, 4$ and hence must satisfy the following equations:

$$n = 1 : 1 = a + b + c + d. \tag{8.2}$$
$$n = 2 : 5 = 8a + 4b + 2c + d. \tag{8.3}$$
$$n = 3 : 14 = 27a + 9b + 3c + d. \tag{8.4}$$
$$n = 4 : 30 = 64a + 16b + 4c + d. \tag{8.5}$$

Subtracting (8.2) from (8.3), (8.3) from (8.4), and (8.4) from (8.5), we then obtain the equations $4 = 7a + 3b + c$, $9 = 19a + 5b + c$, $16 = 37a + 7b + c$. Subtraction of these gives $5 = 12a + 2b$, $7 = 18a + 2b$. Hence we get the solution

$$a = \frac{1}{3}, b = \frac{1}{2}, c = \frac{1}{6}, d = 0.$$

Consequently, our (intelligent) guess is that

$$\sum_{r=1}^{n} r^2 = \frac{1}{3}n^3 + \frac{1}{2}n^2 + \frac{1}{6}n = \frac{1}{6}n(n+1)(2n+1).$$

This turns out to be correct, and we leave it to the reader to prove it by induction. (It is set as Exercise 2 at the end of the chapter in case you forget.)

(Actually, there is a much better way of working out a formula for $\sum_{r=1}^{n} r^2$, given in Exercise 4 at the end of the chapter.)

Geometric Examples

The next example is a nice geometric proof by induction.

Example 8.8
Lines in the plane. If we draw a straight line in the plane, it divides the plane into two regions. If we draw another, not parallel to the first, the two lines divide the plane into four regions. Likewise, three lines, not all going through the same point, and no two of which are parallel, divide the plane into seven regions:

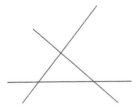

We can carry on drawing lines and counting the regions they form, which leads us naturally to a general question:

If we draw n straight lines in the plane, no three going through the same point, and no two parallel, how many regions do they divide the plane into?

The conditions about not going through the same point and not being parallel may seem strange, but in fact they are very natural: if you draw lines at random, it is very unlikely that two will be parallel or that three will pass through the same point — so you could say the lines in the question are "random" lines. Technically, they are said to be *lines in general position*.

The answers to the question for $n = 1,2,3,4$ are 2,4,7,11. Even from this flimsy evidence you have probably spotted a pattern — the difference between successive terms seems to be increasing by 1 each time. Can we predict a formula from this pattern? Yes, of course we can: the number of regions for one line is two, for two lines is $2+2$, for three lines is $2+2+3$, for four lines is $2+2+3+4$; so we predict that the number of regions for n lines is

$$2+2+3+4+\cdots+n.$$

This is just $1+\Sigma_{r=1}^{n}r$, which by Example 8.6 is equal to $1+\frac{1}{2}n(n+1)$.

Let us therefore attempt to prove the following statement $P(n)$ by induction: the number of regions formed in the plane by n straight lines in general position is $\frac{1}{2}(n^2+n+2)$.

First, $P(1)$ is true, as the number of regions for one line is 2, which is equal to $\frac{1}{2}(1^2+1+2)$.

Now suppose $P(n)$ is true, so n lines in general position form $\frac{1}{2}(n^2+n+2)$ regions. Draw in an $(n+1)^{th}$ line. Since it is not parallel to any of the others, this line meets each of the other n lines in a point, and these n points of intersection divide the $(n+1)^{th}$ line into $n+1$ pieces. Each of these pieces divides an old region into two new ones. Hence, when the $(n+1)^{th}$ line is drawn, the number of regions increases by $n+1$. (If this argument is not clear to you, try drawing a picture to illustrate it when $n=3$ or 4.) Consequently, the number of regions with $n+1$ lines is equal to $\frac{1}{2}(n^2+n+2) + n+1$. Check that this is equal to $\frac{1}{2}((n+1)^2+(n+1)+2)$.

We have now shown that $P(n) \Rightarrow P(n+1)$. Hence, by induction, $P(n)$ is true for all $n \geq 1$.

Induction is a much more powerful method than you might think. It can often be used to prove statements that do not actually explicitly mention an integer n. In such instances, one must imaginatively design a suitable statement $P(n)$ to fit in with the problem and then try to prove $P(n)$ by induction. In the next two examples this is fairly easy to do. The next chapter, however, will be devoted to an example of a proof by induction where the statement $P(n)$ lies a long way away from the initial problem.

Example 8.9

Some straight lines are drawn in the plane, forming regions as in the previous example. Show that it is possible to colour each region either red or blue in such a way that no two neighbouring regions have the same colour.

For example, here is such a colouring when there are three lines:

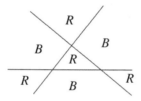

How do we design a suitable statement $P(n)$ for this problem? This is very simple: just take $P(n)$ to be the statement that the regions formed by n straight lines and the plane can be coloured in the required way.

Actually, the proof of $P(n)$ by induction is so neat and elegant that I would hate to deprive you of the pleasure of thinking about it, so I leave it to you. (It is set as Exercise 14 at the end of the chapter in case you forget.)

Prime Factorization

In the next example, we prove a very important result about the integers. First we need a definition:

DEFINITION A prime number *is a positive integer p such that $p \geq 2$ and the only positive integers dividing p are 1 and p.*

You are probably familiar to some extent with prime numbers. The first few are $2, 3, 5, 7, 11, 13, 17, 19, 23, 29, \ldots$.

The important result we shall prove is the following:

PROPOSITION 8.1
Every positive integer greater than 1 is equal to a product of prime numbers.

In the proposition, the number of primes in a product must be allowed to be 1, since a prime number itself is a product of one prime. If n is a positive integer, we call an expression $n = p_1 \ldots p_k$, where $p_1, \ldots, p_k$ are prime numbers, a *prime factorization* of n. Here are some examples of prime factorizations:

$$30 = 2 \cdot 3 \cdot 5, \quad 12 = 2 \cdot 2 \cdot 3, \quad 13 = 13.$$

A suitable statement to attempt to prove by induction is easy to design: for $n \geq 2$, let $P(n)$ be the statement that n is equal to a product of prime numbers.

Clearly $P(2)$ is true, as $2 = 2$ is a prime factorization of 2. However, it is not clear at all how to go about showing that $P(n) \Rightarrow P(n+1)$. In fact this cannot be done, since the primes in the prime factorization of n do not occur in the factorization of $n + 1$.

However, all is not lost. We shall use the following, apparently stronger, principle of induction.

Principle of Strong Mathematical Induction
Suppose that for each integer $n \geq k$ we have a statement $P(n)$. If we prove the following two things:
  (a) $P(k)$ *is true;*
  (b) *for all n, if $P(k), P(k+1), \ldots, P(n)$ are all true, then $P(n+1)$ is also true;*
then $P(n)$ is true for all $n \geq k$.

The logic behind this principle is not really any different from that behind the old principle: by (a), $P(k)$ is true. By (b), $P(k) \Rightarrow P(k+1)$, hence $P(k+1)$ is true. By (b) again, $P(k), P(k+1) \Rightarrow P(k+2)$, hence $P(k+2)$ is true, and so on.

[In fact, the Principle of Strong Induction is actually implied by the old principle. To see this, let $Q(n)$ be the statement that all of $P(k), \ldots, P(n)$ are true. Suppose we have proved (a) and (b) of Strong Induction. Then by (a), $Q(k)$ is true, and by (b), $Q(n) \Rightarrow Q(n+1)$. Hence, by the old principle, $Q(n)$ is true for all $n \geq k$, and therefore so is $P(n)$.]

Let us now apply Strong Induction to prove Proposition 8.1.

Proof of Proposition 8.1   For $n \geq 2$, let $P(n)$ be the statement that n is equal to a product of prime numbers. As we have already remarked, $P(2)$ is true.

Now for part (b) of Strong Induction. Suppose that $P(2), \ldots, P(n)$ are all true. This means that every integer between 2 and n has a prime factorization. Now consider $n+1$. If $n+1$ is prime, then it certainly has a prime factorization (as a product of 1 prime). If $n+1$ is not prime, then by the definition of a prime number, there is an integer a dividing $n+1$ such that $a \neq 1$ or $n+1$. Writing $b = \frac{n+1}{a}$, we then have

$$n+1 = ab \quad \text{and} \quad a, b \in \{2, 3, \ldots, n\}.$$

By assumption, $P(a)$ and $P(b)$ are both true, i.e., a and b have prime factorizations. Say

$$a = p_1 \ldots p_k, \quad b = q_1 \ldots q_l,$$

where all the p_i and q_i are prime numbers. Then

$$n+1 = ab = p_1 \ldots p_k q_1 \ldots q_l.$$

This is an expression for $n+1$ as a product of prime numbers.

We have now shown that $P(2), \ldots, P(n) \Rightarrow P(n+1)$. Therefore, $P(n)$ is true for all $n \geq 2$, by Strong Induction.

Example 8.10
Suppose we are given a sequence of integers $u_0, u_1, u_2, \ldots, u_n, \ldots$ such that $u_0 = 2, u_1 = 3$ and

$$u_{n+1} = 3u_n - 2u_{n-1}$$

for all $n \geq 1$. (Such an equation is called a *recurrence relation* for the sequence.) Can we find a formula for u_n?

Using the equation with $n = 1$, we get $u_2 = 3u_1 - 2u_0 = 5$; and likewise $u_3 = 9, u_4 = 17, u_5 = 33, u_6 = 65$. Is there an obvious pattern? Yes, a reasonable guess seems to be that $u_n = 2^n + 1$.

So let us try to prove by Strong Induction that $u_n = 2^n + 1$. If this is the statement $P(n)$, then $P(0)$ is true, as $u_0 = 2^0 + 1 = 2$. Suppose $P(0), P(1), \ldots, P(n)$ are all true. Then $u_n = 2^n + 1$ and $u_{n-1} = 2^{n-1} + 1$. Hence from the recurrence relation,

$$u_{n+1} = 3\left(2^n + 1\right) - 2\left(2^{n-1} + 1\right) = 3 \cdot 2^n - 2^n + 1 = 2^{n+1} + 1,$$

which shows $P(n+1)$ is true. Therefore, $u_n = 2^n + 1$ for all n, by Strong Induction.

Cauchy's Inequality

This is a famous and important inequality concerning real numbers. Here it is.

PROPOSITION 8.2
Let n be a positive integer. Then for any real numbers $a_1, \ldots, a_n$ and $b_1, \ldots, b_n$,

$$a_1 b_1 + \cdots + a_n b_n \leq \sqrt{a_1^2 + \cdots + a_n^2} \sqrt{b_1^2 + \cdots + b_n^2}.$$

PROOF   Let $P(n)$ be the statement of the proposition. Then $P(1)$ is obvious, and $P(2)$ was proved in Example 5.15.

Assume $P(n)$ is true. Let $a_1, \ldots, a_{n+1}$ and $b_1, \ldots, b_{n+1}$ be real numbers. By $P(n)$,

$$a_1 b_1 + \cdots + a_n b_n \leq \sqrt{a_1^2 + \cdots + a_n^2} \sqrt{b_1^2 + \cdots + b_n^2} = AB,$$

where $A = \sqrt{a_1^2 + \cdots + a_n^2}$, $B = \sqrt{b_1^2 + \cdots + b_n^2}$. Hence

$$a_1 b_1 + \cdots + a_{n+1} b_{n+1} \leq AB + a_{n+1} b_{n+1}.$$

Applying $P(2)$, we have $AB + a_{n+1} b_{n+1} \leq \sqrt{A^2 + a_{n+1}^2} \sqrt{B^2 + b_{n+1}^2}$, and therefore

$$a_1 b_1 + \cdots + a_{n+1} b_{n+1} \leq \sqrt{A^2 + a_{n+1}^2} \sqrt{B^2 + b_{n+1}^2}$$
$$= \sqrt{a_1^2 + \cdots + a_{n+1}^2} \sqrt{b_1^2 + \cdots + b_{n+1}^2}.$$

This is the statement $P(n+1)$. Hence the proposition is proved by induction.   ∎

Cauchy's inequality is applied widely in mathematics. Here are a couple of examples to show how it can be used. Other examples can be found in the Exercises 15–17 at the end of the chapter.

Example 8.11
For any real numbers $a_1, \ldots, a_n$, we have $a_1 + \cdots + a_n \leq \sqrt{n} \sqrt{a_1^2 + \cdots + a_n^2}$.

PROOF   Just apply Proposition 8.2, taking $b_1 = b_2 = \cdots = b_n = 1$.
∎

Example 8.12
Suppose $p_1, \ldots, p_n$ and $a_1, \ldots, a_n$ are real numbers such that $p_i \geq 0$, $a_i \geq 0$ for all i, and $p_1 + \cdots + p_n = 1$. Then

$$(p_1 a_1 + \cdots + p_n a_n)\left(\frac{p_1}{a_1} + \cdots + \frac{p_n}{a_n}\right) \geq 1.$$

PROOF Apply Cauchy's inequality to the sequences $\sqrt{p_1 a_1}, \ldots, \sqrt{p_n a_n}$ and $\sqrt{\frac{p_1}{a_1}}, \ldots, \sqrt{\frac{p_n}{a_n}}$. ∎

Exercises for Chapter 8

1. Prove by induction that it is possible to pay, without requiring change, any whole number of roubles greater than 7 with banknotes of value 3 roubles and 5 roubles.

2. Prove by induction that $\Sigma_{r=1}^n r^2 = \frac{1}{6}n(n+1)(2n+1)$.

   Deduce formulae for

   $$1 \cdot 1 + 2 \cdot 3 + 3 \cdot 5 + 4 \cdot 7 + \cdots + n(2n-1) \quad \text{and} \quad 1^2 + 3^2 + 5^2 + \cdots (2n-1)^2.$$

3. (a) Work out 1, $1+8$, $1+8+27$ and $1+8+27+64$. Guess a formula for $\Sigma_{r=1}^n r^3$ and prove it.

   (b) Check that $1 = 0+1$, $2+3+4 = 1+8$ and $5+6+\cdots+9 = 8+27$. Find a general formula for which these are the first three cases. Prove your formula is correct.

4. Here is another way to work out $\Sigma_{r=1}^n r^2$. Observe that $(r+1)^3 - r^3 = 3r^2 + 3r + 1$. Hence

   $$\sum_{r=1}^n (r+1)^3 - r^3 = 3\sum_{r=1}^n r^2 + 3\sum_{r=1}^n r + n.$$

   The left-hand side is equal to

   $$\left(2^3 - 1^3\right) + \left(3^3 - 2^3\right) + \left(4^3 - 3^3\right) + \cdots$$
   $$+ \left((n+1)^3 - n^3\right) = (n+1)^3 - 1.$$

   Hence we can work out $\Sigma_{r=1}^n r^2$.

   Carry out this calculation, and check that your formula agrees with that in Exercise 2.

   Use the same method to work out formulae for $\Sigma_{r=1}^n r^3$ and $\Sigma_{r=1}^n r^4$.

5. Prove the following statements by induction:

    (a) For all integers $n \geq 0$, the number $5^{2n} - 3^n$ is a multiple of 11.

    (b) For any integer $n \geq 1$, the integer 2^{4n-1} ends with an 8.

    (c) The sum of the cubes of three consecutive positive integers is always a multiple of 9.

    (d) If $x \geq 2$ is a real number and $n \geq 1$ is an integer, then $x^n \geq nx$.

    (e) If $n \geq 3$ is an integer, then $5^n > 4^n + 3^n + 2^n$.

6. The *Lucas sequence* is a sequence of integers $l_1, l_2, \ldots, l_n, \ldots$, such that $l_1 = 1, l_2 = 3$ and
$$l_{n+1} = l_n + l_{n-1}$$
    for all $n \geq 1$. So the sequence starts $1, 3, 4, 7, 11, 18, \ldots$

    Find the pattern for the remainders when l_n is divided by 3. (*Hint:* Consider the first 8 remainders, then the next 8, and so on; formulate a conjecture for the pattern and prove it by induction.)

    Is L_{2000} divisible by 3?

7. The *Fibonacci sequence* is a sequence of integers $f_1, f_2, \ldots, f_n, \ldots$, such that $f_1 = 1, f_2 = 1$ and

$$f_{n+1} = f_n + f_{n-1}$$

    for all $n \geq 1$. Prove by strong induction that for all n,

$$f_n = \frac{1}{\sqrt{5}} (\alpha^n - \beta^n),$$

    where $\alpha = \frac{1+\sqrt{5}}{2}$ and $\beta = \frac{1-\sqrt{5}}{2}$.

8. I just worked out $(2 + \sqrt{3})^{50}$ on my computer and got the answer

    395710319992261395631627353 73.99999999999999999999999999974728...

    Why is this so close to an integer?

    (*Hint:* Try to use the idea of the previous question by constructing a suitable sequence.)

9. Prove that if $0 < q < \frac{1}{2}$, then for all $n \geq 1$,

$$(1+q)^n \leq 1 + 2^n q.$$

10. (a) Prove that for every integer $n \geq 2$,

$$\frac{1}{n+1} + \frac{1}{n+2} + \cdots + \frac{1}{2n} \geq \frac{7}{12}.$$

(b) Prove that for every integer $n \geq 1$,

$$1 + \frac{1}{\sqrt{2}} + \frac{1}{\sqrt{3}} + \cdots + \frac{1}{\sqrt{n}} \leq 2\sqrt{n}.$$

11. Just for this question, count 1 as a prime number. A well-known result in number theory says that for every integer $x \geq 3$, there is a prime number p such that $\frac{1}{2}x < p < x$. Using this result and strong induction, prove that every positive integer is equal to a sum of primes, all of which are different.

12. Here is a "proof" by induction that any two positive integers are equal (e.g., $5 = 10$):

First, a definition: if a and b are positive integers, define $\max(a,b)$ to be the larger of a and b if $a \neq b$, and to be a if $a = b$. [For instance, $\max(3,5) = 5$, $\max(3,3) = 3$.] Let $P(n)$ be the statement: "if a and b are positive integers such that $\max(a,b) = n$, then $a = b$." We prove $P(n)$ true for all $n \geq 1$ by induction. [As a consequence, if a,b are any two positive integers, then $a = b$, since $P(n)$ is true, where $n = \max(a,b)$.]

First, $P(1)$ is true, since if $\max(a,b) = 1$ then a and b must both be equal to 1. Now assume $P(n)$ is true. Let a,b be positive integers such that $\max(a,b) = n+1$. Then $\max(a-1,b-1) = n$. As we are assuming $P(n)$, this implies that $a-1 = b-1$, hence $a = b$. Therefore, $P(n+1)$ is true. By induction, $P(n)$ is true for all n.

There must be something wrong with this "proof." Can you find the error?

13. (a) Suppose we have n straight lines in a plane, and all the lines pass through a single point. Into how many regions do the lines divide the plane? Prove your answer.

(b) We know from Example 8.8 that n straight lines in general position in a plane divide the plane into $\frac{1}{2}(n^2 + n + 2)$ regions. How many of these regions are infinite and how many are finite?

(In case of any confusion, a finite region is one that has finite area; an infinite region is one that does not.)

14. (See Example 8.9.) Some straight lines are drawn in the plane, forming regions. Show that it is possible to colour each region either red or blue in such a way that no two neighbouring regions have the same colour.

15. (a) Prove that $a^2 + b^2 + c^2 \geq ab + bc + ca$ for all real numbers a, b, c.

    (b) Suppose x, y, z are positive and $x + y + z = 1$. Prove that

        (i) $x^2 + y^2 + z^2 \geq \frac{1}{3}$, and

    (ii) $\frac{1}{x} + \frac{1}{y} + \frac{1}{z} \geq 9$.

16. Let x, y, z be positive real numbers. Prove the following:

    (i) $\sqrt{\frac{x+y}{x+y+z}} + \sqrt{\frac{x+z}{x+y+z}} + \sqrt{\frac{y+z}{x+y+z}} \leq \sqrt{6}$,

    (ii) $\frac{x^2}{y+z} + \frac{y^2}{x+z} + \frac{z^2}{x+y} \geq \frac{1}{2}(x+y+z)$.

    In both parts, give examples to show that equality can hold.

17. Let $p_1, \ldots, p_n$ be positive real numbers such that $p_1 + \cdots + p_n = 1$.

    (a) Prove the following:

        (i) $\sum_{i=1}^{n} p_i^2 \geq \frac{1}{n}$.

        (ii) $\sum_{i=1}^{n} \frac{1}{p_i} \geq n^2$.

        (iii) $\sum_{i=1}^{n} \frac{1}{p_i^2} \geq n^3$.

    (b) Deduce that

$$\sum_{i=1}^{n} \left(p_i + \frac{1}{p_i} \right)^2 \geq \frac{(n^2+1)^2}{n}.$$

18. Critic Ivor Smallbrain is sitting through a showing of the new film *Poly-gon with the Wind*. Ivor is not enjoying the film and begins to doodle on a piece of paper, drawing circles in such a way that any two of the circles intersect, no two circles touch each other, and no three circles pass through the same point. He notices that after drawing two circles he has divided the plane into four regions, after three there are eight regions, and he wonders to himself how many regions there will be after he has drawn n circles. Can you help Ivor?

Chapter 9

Euler's Formula and Platonic Solids

This chapter contains a rather spectacular proof by induction. The result we shall prove is a famous formula of Euler from the eighteenth century, concerning the relationship between the numbers of corners, edges and faces of a solid object. As an application of Euler's formula we shall then study the five Platonic solids — the cube, regular tetrahedron, octahedron, icosahedron and dodecahedron.

We shall call our solid objects *polyhedra*. A *polyhedron* is a solid whose surface consists of a number of faces, all of which are polygons, such that any side of a face lies on exactly one other face. The corners of the faces are called the *vertices* of the polyhedron, and their sides are the *edges* of the polyhedron.

Here are some everyday examples of polyhedra.

(1) *Cube*

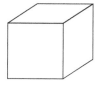

This has 8 vertices, 12 edges and 6 faces.

(2) *Tetrahedron*

This has 4 vertices, 6 edges and 4 faces.

(3) *Triangular prism*

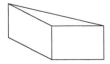

This has 6 vertices, 9 edges and 5 faces.

(4) *n-prism* This is like the triangular prism, except that its top and bottom faces are n-sided polygons rather than triangles. It has $2n$ vertices, $3n$ edges and $n + 2$ faces.

Let us collect the numbers of vertices, edges and faces for the above examples in a table. Denote these numbers by V, E and F, respectively.

	V	E	F
(1)	8	12	6
(2)	4	6	4
(3)	6	9	5
(4)	$2n$	$3n$	$n+2$

Can you see a relationship between these numbers that holds in every case? You probably can — it is

$$V - E + F = 2.$$

This is Euler's famous formula, and we shall show that it holds in general for all *convex* polyhedra: a polyhedron is convex if, whenever we choose two points on its surface, the straight line joining them lies entirely within the polyhedron.

All of the above examples are convex polyhedra. However, if we for example take a cube and remove a smaller cube from its interior, we get a polyhedron that is not convex; for this polyhedron, in fact, $V = 16, E = 24, F = 12$, so $V - E + F = 4$ and the formula fails.

Here then is Euler's formula.

THEOREM 9.1

For a convex polyhedron with V vertices, E edges and F faces, we have

$$V - E + F = 2.$$

As I said, we shall prove this result by induction. So somehow we have to design a suitable statement $P(n)$ to try to prove by induction. What on earth should $P(n)$ be?

Before going into this, let us first translate the problem from one about objects in 3-dimensional space to one about objects in the plane. Take a convex polyhedron as in the theorem, and choose one face of it. Regard this face as a window, put your eye very close to the window, and draw on the window pane the vertices and edges you can see through the window. The result is a figure in the plane with straight edges, vertices and faces. For example, here is what we would draw for the cube, the tetrahedron and the triangular prism:

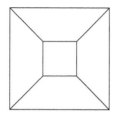  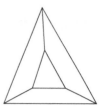

(The outer edges enclose the window.)

The resulting figure in the plane has V vertices, E edges and $F - 1$ faces (we lose one face, since the window is no longer a face). It is a "connected plane graph," in the sense of the following definition.

DEFINITION   *A plane graph is a figure in the plane consisting of a collection of points (vertices), and some edges joining various pairs of these points, with no two edges crossing each other. A plane graph is connected if we can get from any vertex of the graph to any other vertex by going along a path of edges in the graph.*

For example, here is a connected plane graph:

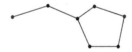

It has 7 vertices, 7 edges and 1 face.

THEOREM 9.2
If a connected plane graph has v vertices, e edges and f faces, then

$$v - e + f = 1.$$

Theorem 9.2 easily implies Euler's Theorem 9.1: for if we have a convex polyhedron with V vertices, E edges and F faces, then as explained above we get a connected plane graph with V vertices, E edges and $F - 1$ faces. If we knew Theorem 9.2 was true, we could then deduce that $V - E + (F - 1) = 1$, hence $V - E + F = 2$, as required for Euler's theorem.

So we need to prove Theorem 9.2.

Proof of Theorem 9.2   Here is the statement $P(n)$ that we are going to try
to prove by induction:

$P(n)$: every connected plane graph with n edges satisfies the formula
$v - n + f = 1$.

Notice that $P(n)$ is a statement about lots of plane graphs. $P(1)$ says that every
connected plane graph with 1 edge satisfies the formula; there is only one such
graph:

This has 2 vertices, 1 edge and 0 faces, and since $2 - 1 + 0 = 1$, it satisfies the
formula. Likewise, $P(2)$ says that this graph

satisfies the formula, which it does, as $v = 3, e = 2, f = 0$. For $P(3)$, there are
three different connected plane graphs with 3 edges:

Each satisfies the formula.

Let us prove $P(n)$ by induction. First, $P(1)$ is true, as observed in the previous paragraph.

Now assume $P(n)$ is true — so every connected plane graph with n edges
satisfies the formula. We need to deduce $P(n + 1)$. So consider a connected
plane graph G with $n + 1$ edges. Say G has v vertices and f faces. We want to
prove that G satisfies the formula $v - (n + 1) + f = 1$.

Our strategy will be to remove a carefully chosen edge from G, so as to leave
a connected plane graph with only n edges, and then use $P(n)$.

If G has at least 1 face (i.e., $f \geq 1$), we remove one edge of this face. The
remaining graph G' is still connected and has n edges, v vertices and $f - 1$
faces. Since we are assuming $P(n)$, we know that G' satisfies the formula,
hence

$$v - n + (f - 1) = 1.$$

Therefore $v - (n + 1) + f = 1$, as required.

If G has no faces at all (i.e., $f = 0$), then it has at least one "end-vertex" —
that is, a vertex that is joined by an edge to only one other vertex (see Exercise
5 at the end of the chapter). Removing this end-vertex and its edge from G
leaves a connected plane graph G'' with $v - 1$ vertices, n edges and 0 faces. By
$P(n)$, G'' satisfies the formula, so

$$(v - 1) - n + 0 = 1.$$

Hence $v - (n+1) + 0 = 1$, which is the formula for G.

We have established that $P(n) \Rightarrow P(n+1)$, so $P(n)$ is true for all n by induction.

This completes the proof of Theorem 9.2, and hence of Euler's Theorem 9.1.

Regular and Platonic Solids

A polygon is said to be *regular* if all its sides are of equal length and all its internal angles are equal. We call a regular polygon with n sides a *regular n-gon*. Some of these shapes are probably quite familiar; for example, a regular n-gon with $n = 3$ is just an equilateral triangle, $n = 4$ is a square, $n = 5$ is a regular pentagon, and so on:

A polyhedron is *regular* if its faces are regular polygons, all with the same number of sides, and also each vertex belongs to the same number of edges.

Three examples of regular polyhedra come more or less readily to mind: the cube, the tetrahedron and the octahedron. These are three of the famous five *Platonic solids;* the other two are the less obvious *icosahedron,* which has 20 triangular faces, and *dodecahedron,* which has 12 pentagonal faces. Here are pictures of the octahedron, icosahedron and dodecahedron:

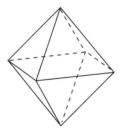 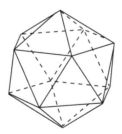 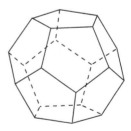

Every regular polyhedron carries five associated numbers: three are V, E, F, and the other two are n, the number of sides on a face, and r, the number of edges each vertex belongs to. We record these numbers for the Platonic solids:

	V	E	F	n	r
tetrahedron	4	6	4	3	3
cube	8	12	6	4	3
octahedron	6	12	8	3	4
icosahedron	12	30	20	3	5
dodecahedron	20	30	12	5	3

As you might have guessed from the name, the Platonic solids were known to the Greeks. They are the most symmetrical, elegant and robust of solids, so it is natural to look for more regular polyhedra. Remarkably, though perhaps disappointingly, there are no others. This fact is another theorem of the great Euler. The proof is a wonderful application of Euler's formula 9.1. Here it is.

THEOREM 9.3
The only regular convex polyhedra are the five Platonic solids.

PROOF   Suppose we have a regular polyhedron with parameters V, E, F, n and r as defined above.

First we need to show some relationships between these parameters. We shall prove first that

$$2E = nF. \tag{9.1}$$

To prove this, let us calculate the number of pairs

$$e, f$$

where e is an edge, f is a face and e lies on f. Well, there are E possibilities for the edge e, and each lies in 2 faces f; so the number of such pairs e, f is equal to $2E$. On the other hand, there are F possibilities for the face f, and each has n edges e; so the number of such pairs e, f is also equal to nF. Therefore, $2E = nF$, proving (9.1).

Next we show that

$$2E = rV. \tag{9.2}$$

The proof of this is quite similar: count the pairs

$$v, e$$

where v is a vertex, e an edge and v lies on e. There are E edges e, and each has 2 vertices v, so the number of such pairs v, e is $2E$; on the other hand, there are V vertices v, and each lies on r edges, so the number of such pairs is also rV. This proves (9.2).

At this point we appeal to Euler's formula 9.1:

$$V - E + F = 2.$$

Substituting $V = \frac{2E}{r}, F = \frac{2E}{n}$ from (9.1) and (9.2), we obtain $\frac{2E}{r} - E + \frac{2E}{n} = 2$; hence

$$\frac{1}{r} + \frac{1}{n} = \frac{1}{2} + \frac{1}{E}. \tag{9.3}$$

Now we know that $n \geq 3$, as a polygon must have at least 3 sides; likewise $r \geq 3$, since it is geometrically clear that in a polyhedron a vertex must belong to at least 3 edges. By (9.3), it certainly cannot be the case that both $n \geq 4$ and $r \geq 4$, since this would make the left-hand side of (9.3) at most $\frac{1}{2}$, whereas the right-hand side is more than $\frac{1}{2}$. It follows that either $n = 3$ or $r = 3$.

If $n = 3$, then (9.3) becomes

$$\frac{1}{r} = \frac{1}{6} + \frac{1}{E}.$$

The right-hand side is greater than $\frac{1}{6}$, and hence $r < 6$. Therefore, $r = 3$, 4 or 5 and $E = 6$, 12 or 30, respectively. The possible values of V, F are given by (9.1) and (9.2).

Likewise, if $r = 3$, (9.3) becomes $\frac{1}{n} = \frac{1}{6} + \frac{1}{E}$, and we argue similarly that $n = 3, 4$ or 5 and $E = 6, 12$ or 30, respectively.

We have now shown that the numbers V, E, F, n, r for a regular polyhedron must be one of the possibilities in the following table:

V	E	F	n	r
4	6	4	3	3
8	12	6	4	3
6	12	8	3	4
12	30	20	3	5
20	30	12	5	3

These are the parameter sets of the tetrahedron, cube, octahedron, icosahedron and dodecahedron, respectively. To complete the proof we now only have to show that each Platonic solid is the only regular solid with its particular parameter set. This is a simple geometric argument, and we present it just for the last parameter set — the proofs for the other sets are entirely similar.

So suppose we have a regular polyhedron with 20 pentagonal faces, each vertex lying on 3 edges. Focus on a particular vertex. At this vertex there is only one way of fitting three pentagonal faces together:

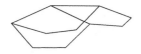

At each of the other vertices of these three pentagons, there is likewise only one way of fitting two further pentagons together. Carrying on this argument with all new vertices, we see that there is at most one way to make a regular solid with these parameters. Since the dodecahedron is such a solid, it is the only one. This completes the proof. ∎

Exercises for Chapter 9

1. Consider a convex polyhedron, all of whose faces are squares or regular pentagons. Say there are m squares and n pentagons. Assume that each vertex lies on exactly 3 edges.

   (a) Show that for this polyhedron, the following equations hold:

   $$3V = 2E, \quad 4m+5n = 2E, \quad m+n = F.$$

   (b) Using Euler's formula, deduce that $2m+n = 12$.

   (c) Find examples of such polyhedra for as many different values of m as you can.

2. Prove that for a convex polyhedron with V vertices, E edges and F faces, the following inequalities are true:

   $$2E \geq 3F \quad \text{and} \quad 2E \geq 3V.$$

   Deduce using Euler's formula that

   $$2V \geq F+4, \quad 3V \geq E+6, \quad 2F \geq V+4 \quad \text{and} \quad 3F \geq E+6.$$

   Give an example of a convex polyhedron for which all these inequalities are equalities (i.e., $2V = F+4$, etc.).

3. Prove that if a connected plane graph has v vertices and e edges, and $v \geq 3$, then $e \leq 3v-6$.

4. Prove that it is impossible to make a football out of exactly 9 squares and m octagons, where $m \geq 4$. (In this context, a "football" is a convex polyhedron in which at least 3 edges meet at each vertex.)

5. Prove that if a finite connected plane graph has no faces, then it has a vertex that is joined to exactly one other vertex. (*Hint:* Assume for a contradiction that every vertex is joined to at least two others. Try to use this to show there must be a face.)

6. Draw all the connected plane graphs with 4 edges, and all the connected plane graphs with 4 vertices.

7. Let K_n denote the graph with n vertices in which any two vertices are joined by an edge. So, for example, K_2 consists of 2 vertices joined by an edge and K_3 is a triangle.

   Prove that it is possible to draw K_4 as a plane graph.

   Prove that it is impossible to draw K_5 as a plane graph. (*Hint:* Use the inequality in Exercise 3 cleverly.)

8. Prove that every connected plane graph has a vertex that is joined to at most five other vertices. (*Hint:* Assume every vertex is joined to at least 6 others, and try to use Exercise 3 to get a contradiction.)

9. Critic Ivor Smallbrain has been thrown into prison for libelling the great film director Michael Loser. During one of his needlework classes in prison, Ivor is given a pile of pieces of leather in the shapes of regular pentagons and regular hexagons and is told to sew some of these together into a convex polyhedron (which will then be used as a football). He is told that each vertex must lie on exactly 3 edges. Ivor immediately exclaims, "Then I need exactly 12 pentagonal pieces!"

   Prove that Ivor is correct.

Chapter 10

The Integers

In this chapter we begin to study the most basic, and also perhaps the most fascinating, number system of all — the integers. Our first aim will be to investigate factorization properties of integers. We know already that every integer greater than 1 has a prime factorization (Proposition 8.1). This was quite easy to prove using Strong Induction. A somewhat more delicate question is whether the prime factorization of an integer is always *unique* — in other words, whether, given an integer n, one can write it as a product of primes in only one way. The answer is yes; and this is such an important result that it has acquired the grandiose title of "The Fundamental Theorem of Arithmetic." We shall prove it in the next chapter and try there to show why it is such an important result by giving some examples of its use. In this chapter we lay the groundwork for this.

We begin with a familiar definition.

DEFINITION Let $a, b \in \mathbb{Z}$. *We say a divides b (or a is a factor of b) if $b = ac$ for some integer c. When a divides b, we write $a|b$.*

Usually, of course, given two integers a, b at random, it is unlikely that a will divide b. But we can "divide a into b" and get a quotient and a remainder:

PROPOSITION 10.1
Let a be a positive integer. Then for any $b \in \mathbb{Z}$, there are integers q, r such that
$$b = qa + r \quad and \quad 0 \leq r < a.$$

The integer q is called the quotient, and r is the remainder. For example, if $a = 17, b = 183$ then the equation in Proposition 10.1 is $183 = 10 \cdot 17 + 13$, the quotient is 10 and the remainder 13.

PROOF Consider the rational number $\frac{b}{a}$. There is an integer q such that

$$q \le \frac{b}{a} < q+1$$

(this is just saying $\frac{b}{a}$ lies between two consecutive integers). Multiplying through by the positive integer a, we obtain $qa \le b < (q+1)a$, hence $0 \le b - qa < a$.

Now put $r = b - qa$. Then $b = qa + r$ and $0 \le r < a$, as required. ∎

PROPOSITION 10.2
Let $a,b,d \in \mathbb{Z}$, and suppose that $d|a$ and $d|b$. Then $d|(ma+nb)$ for any $m,n \in \mathbb{Z}$.

PROOF   Let $a = c_1 d$ and $b = c_2 d$ with $c_1, c_2 \in \mathbb{Z}$. Then for $m,n \in \mathbb{Z}$,

$$ma + nb = mc_1 d + nc_2 d = (mc_1 + nc_2)d.$$

Hence $d|(ma+nb)$. ∎

The Euclidean Algorithm

The Euclidean algorithm is a step-by-step method for calculating the common factors of two integers. First we need a definition.

DEFINITION   *Let $a,b \in \mathbb{Z}$. A* common factor *of a and b is an integer that divides both a and b. The* highest common factor *of a and b, written $hcf(a,b)$, is the largest positive integer that divides both a and b.*

For example, $hcf(2,3) = 1$ and $hcf(4,6) = 2$. But how do we go about finding the highest common factor of two large numbers, say 5817 and 1428? This is what the Euclidean algorithm does for us — in a few simple, mindless steps.

Before presenting the algorithm in all its full glory, let us do an example.

Example 10.1
Here we find $hcf(5817, 1428)$ in a few mindless steps, as advertised. Write $b = 5817, a = 1428$, and let $d = hcf(a,b)$.

Step 1 Divide a into b and get a quotient and remainder:

$$5817 = 4 \cdot 1428 + 105.$$

(*Deduction*: As $d|a$ and $d|b$, d also divides $a - 4b = 105$.)

Step 2 Divide 105 into 1428:

$$1428 = 13 \cdot 105 + 63.$$

(*Deduction*: As $d|1428$ and $d|105$, d also divides 63.)

Step 3 Divide 63 into 105:

$$105 = 1 \cdot 63 + 42.$$

(*Deduction*: $d|42$.)

Step 4 Divide 42 into 63:

$$63 = 1 \cdot 42 + 21.$$

(*Deduction*: $d|21$.)

Step 5 Divide 21 into 42:

$$42 = 2 \cdot 21 + 0.$$

Step 6 STOP!

We claim that $d = \text{hcf}(5817, 1428) = 21$, the last non-zero remainder in the above steps. We have already observed that $d|21$. To prove our claim, we work upwards from the last step to the first: namely, Step 5 shows that $21|42$; hence Step 4 shows that $21|63$; hence Step 3 shows that $21|105$; hence Step 2 shows that $21|1428$; hence Step 1 shows $21|5817$. Therefore, 21 divides both a and b, so $d \geq 21$. As $d|21$, it follows that $d = 21$, as claimed.

The general version of the Euclidean algorithm is really no more complicated than this example. Here it is.

Let a, b be integers. To calculate $\text{hcf}(a, b)$, we perform (mindless) steps as in the example: first divide a into b, getting a quotient q_1 and remainder r_1; then divide r_1 into a, getting remainder $r_2 < r_1$; then divide r_2 into r_1, getting remainder $r_3 < r_2$; and carry on like this until we eventually get a remainder 0 (which we must, as the r_is are decreasing and are ≥ 0). Say the remainder 0 occurs after $n+1$ steps. Then the equations representing the steps are:

$$(1)\ b = q_1 a + r_1 \qquad\qquad \text{with } 0 \le r_1 < a$$
$$(2)\ a = q_2 r_1 + r_2 \qquad\qquad \text{with } 0 \le r_2 < r_1$$
$$(3)\ r_1 = q_3 r_2 + r_3 \qquad\qquad \text{with } 0 \le r_3 < r_2$$
$$\cdots \qquad\qquad\qquad\qquad \cdot$$
$$\cdots \qquad\qquad\qquad\qquad \cdot$$
$$\cdots \qquad\qquad\qquad\qquad \cdot$$
$$(n-1)\ r_{n-3} = q_{n-1} r_{n-2} + r_{n-1} \quad \text{with } 0 \le r_{n-1} < r_{n-2}$$
$$(n)\ r_{n-2} = q_n r_{n-1} + r_n \qquad\quad \text{with } 0 \le r_n < r_{n-1}$$
$$(n+1)\ r_{n-1} = q_{n+1} r_n + 0.$$

THEOREM 10.1

In the above, the highest common factor hcf(a,b) *is equal to* r_n, *the last non-zero remainder.*

PROOF   Let $d = $ hcf(a,b). We first show that $d|r_n$ by arguing from equation (1) downwards. By Proposition 10.2, d divides $b - q_1 a$, and hence by (1), $d|r_1$. Then by (2), $d|r_2$; by (3), $d|r_3$; and so on, until by (n), $d|r_n$.

Now we show that $d \ge r_n$ by working upwards from equation $(n+1)$. By $(n+1)$, $r_n | r_{n-1}$; hence by (n), $r_n | r_{n-2}$; hence by $(n-1)$, $r_n | r_{n-3}$; and so on, until by (2), $r_n | a$ and then by (1), $r_n | b$. Thus, r_n is a common factor of a and b, and so $d \ge r_n$.

We conclude that $d = r_n$, and the proof is complete.   ∎

The next result is an important consequence of the Euclidean algorithm.

PROPOSITION 10.3

If $a, b \in \mathbb{Z}$ *and* $d = $ hcf(a,b), *then there are integers* s *and* t *such that*

$$d = sa + tb.$$

PROOF   We use Equations (1),..., (n) above. By (n),

$$d = r_n = r_{n-2} - q_n r_{n-1}.$$

Substituting for r_{n-1} using Equation $(n-1)$, we get

$$d = r_{n-2} - q_n(r_{n-3} - q_{n-1} r_{n-2}) = x r_{n-2} + y r_{n-3}$$

where $x, y \in \mathbb{Z}$. Using Equation $(n-2)$, we can substitute for r_{n-2} in this (specifically, $r_{n-2} = r_{n-4} - q_{n-2} r_{n-3}$), to get

$$d = x' r_{n-3} + y' r_{n-4}$$

where $x', y' \in \mathbb{Z}$. Carrying on like this, we eventually get $d = sa + tb$ with $s, t \in \mathbb{Z}$, as required. ∎

Example 10.2

We know by Example 10.1 that $\mathrm{hcf}(5817, 1428) = 21$. So by Proposition 10.3 there are integers s, t such that

$$21 = 5817s + 1428t.$$

Let us find such integers s, t.

To do this, we apply the method given in the proof of Proposition 10.3, using the equations in Steps 1 through 4 of Example 10.1. By Step 4,

$$21 = 63 - 42.$$

Hence by Step 3,

$$21 = 63 - (105 - 63) = -105 + 2 \cdot 63.$$

Hence by Step 2,

$$21 = -105 + 2(1428 - 13 \cdot 105) = 2 \cdot 1428 - 27 \cdot 105.$$

Hence by Step 1,

$$21 = 2 \cdot 1428 - 27(5817 - 4 \cdot 1428) = -27 \cdot 5817 + 110 \cdot 1428.$$

Thus we have found our integers s, t: $s = -27, t = 110$ will work. (But note that there are many other values of s, t which also work; for example, $s = -27 + 1428, t = 110 - 5817$.)

Here is a consequence of Proposition 10.3.

PROPOSITION 10.4

If $a, b \in \mathbb{Z}$, then any common factor of a and b also divides $\mathrm{hcf}(a, b)$.

PROOF   Let $d = \mathrm{hcf}(a, b)$. By Proposition 10.3, there are integers s, t such that $d = sa + tb$. If m is a common factor of a and b, then m divides $sa + tb$ by Proposition 10.2, and hence m divides d. ∎

We are now in a position to prove a highly significant fact about prime numbers: namely, that if a prime number p divides a product ab of two integers, then p divides one of the factors a and b.

DEFINITION　　*If $a,b \in \mathbb{Z}$ and $hcf(a,b) = 1$, we say that a and b are* coprime *to each other.*

For example, 17 and 1024 are coprime to each other. Note that by Proposition 10.3, if a,b are coprime to each other, then there are integers s,t such that $1 = sa + tb$.

PROPOSITION 10.5

Let $a,b \in \mathbb{Z}$.

(a) *Suppose c is an integer such that a,c are coprime to each other, and $c|ab$. Then $c|b$.*

(b) *Suppose p is a prime number and $p|ab$. Then either $p|a$ or $p|b$.*

PROOF　　(a) By Proposition 10.3, there are integers s,t such that $1 = sa + tc$. Multiplying through by b gives

$$b = sab + tcb.$$

Since $c|ab$ and $c|tcb$, the right-hand side is divisible by c. Hence $c|b$.

(b) We show that if p does not divide a, then $p|b$. Suppose then that p does not divide a. As the only positive integers dividing p are 1 and p, $hcf(a,p)$ must be 1 or p. It is not p as p does not divide a; hence $hcf(a,p) = 1$. Thus a,p are coprime to each other and $p|ab$. It follows by part (a) that $p|b$, as required.　■

Proposition 10.5(b) will be crucial in our proof of the uniqueness of prime factorization in the next chapter. To apply it there, we need to generalize it slightly to the case of a prime dividing a product of many factors, as follows.

PROPOSITION 10.6

Let $a_1, a_2, \ldots, a_n \in \mathbb{Z}$, and let p be a prime number. If $p|a_1a_2\ldots a_n$, then $p|a_i$ for some i.

PROOF　　We prove this by induction. Let $P(n)$ be the statement of the proposition.

First, $P(1)$ says "if $p|a_1$ then $p|a_1$," which is trivially true.

Now suppose $P(n)$ is true. Let $a_1, a_2, \ldots, a_{n+1} \in \mathbb{Z}$, with $p|a_1a_2\ldots a_{n+1}$. We need to show that $p|a_i$ for some i.

Regard $a_1a_2\ldots a_{n+1}$ as a product ab, where $a = a_1a_2\ldots a_n$ and $b = a_{n+1}$. Then $p|ab$, so by Proposition 10.5(b), either $p|a$ or $p|b$. If $p|a$, that is to say $p|a_1a_2, \ldots, a_n$, then by $P(n)$ we have $p|a_i$ for some i; and if $p|b$ then $p|a_{n+1}$. Thus, in either case, p divides one of the factors $a_1, a_2, \ldots, a_{n+1}$.

We have established that $P(n) \Rightarrow P(n+1)$. Hence, by induction, $P(n)$ is true for all n. ∎

Exercises for Chapter 10

1. For each of the following pairs a,b of integers, find the highest common factor $d = \mathrm{hcf}(a,b)$, and find integers s,t such that $d = sa + tb$:

   (i) $a = 17, b = 29$.

   (ii) $a = 552, b = 713$.

   (iii) $a = 345, b = 299$.

2. Show that if a,b are positive integers and $d = \mathrm{hcf}(a,b)$, then there are positive integers s,t such that $d = sa - tb$.

   Find such positive integers s,t in each of cases (i)–(iii) in Exercise 1.

3. A train leaves Moscow for St. Petersburg every 7 hours, on the hour. Show that on some days it is possible to catch this train at 9 a.m.

   Whenever there is a 9 a.m. train, Ivan takes it to visit his aunt Olga. How often does Olga see her nephew?

   Discuss the corresponding problem involving the train to Vladivostok, which leaves Moscow every 14 hours.

4. (a) Show that for all positive integers n, $\mathrm{hcf}(6n+8, 4n+5) = 1$.

   (b) Suppose a,b are integers such that $a|b$ and $b|a$. Prove that $a = \pm b$.

   (c) Suppose s,t,a,b are integers such that $sa + tb = 1$. Show that $\mathrm{hcf}(a,b) = 1$.

5. (a) Let m,n be coprime integers, and suppose a is an integer which is divisible by both m and n. Prove that mn divides a.

   (b) Show that the conclusion of part (a) is false if m and n are not coprime (i.e., show that if m and n are not coprime, there exists an integer a such that $m|a$ and $n|a$, but mn does not divide a).

   (c) Show that if $\mathrm{hcf}(x,m) = 1$ and $\mathrm{hcf}(y,m) = 1$, then $\mathrm{hcf}(xy,m) = 1$.

6. Let $a,b,c \in \mathbb{Z}$. Define the highest common factor $\mathrm{hcf}(a,b,c)$ to be the largest positive integer that divides a,b and c. Prove that there are integers s,t,u such that

   $$\mathrm{hcf}(a,b,c) = sa + tb + uc.$$

   Find such integers s,t,u when $a = 91, b = 903, c = 1792$.

7. Jim plays chess every 3 days, and his friend Marmaduke eats spaghetti every 4 days. One Sunday it happens that Jim plays chess and Marmaduke eats spaghetti. How long will it be before this again happens on a Sunday?

8. Let $n \geq 2$ be an integer. Prove that n is prime if and only if for every integer a, either $\text{hcf}(a,n) = 1$ or $n|a$.

9. Let a,b be coprime positive integers. Prove that for any integer n there exist integers s,t with $s > 0$ such that $sa + tb = n$.

10. After a particularly exciting viewing of the new Danish thriller *Den hele tal*, critic Ivor Smallbrain repairs for refreshment to the prison's high-security fast-food outlet O'Ducks. He decides that he'd like to eat some delicious Chicken O'Nuggets. These are sold in packs of two sizes — one containing 4 O'Nuggets, and the other containing 9 O'Nuggets.

    Prove that for any integer $n > 23$, it is possible for Ivor to buy n O'Nuggets (assuming he has enough money).

    Perversely, however, Ivor decides that he must buy exactly 23 O'Nuggets, no more and no less. Is he able to do this?

    Generalize this question, replacing 4 and 9 by any pair a,b of coprime positive integers: find an integer N (depending on a and b), such that for any integer $n > N$ it is possible to find integers $s,t \geq 0$ satisfying $sa + tb = n$, but no such s,t exist satisfying $sa + tb = N$.

Chapter 11

Prime Factorization

We have already seen in Chapter 8 (Proposition 8.1) that every integer greater than 1 is equal to a product of prime numbers; that is, it has a prime factorization. The main result of this chapter, the Fundamental Theorem of Arithmetic, tells us that this prime factorization is unique — in other words, there is essentially only one way of writing an integer as a product of primes. (In case you think this is somehow obvious, have a look at Exercise 6 at the end of the chapter to find an example of a number system where prime factorization is *not* unique.)

The Fundamental Theorem of Arithmetic may not seem terribly thrilling to you at first sight. However, it is in fact one of the most important properties of the integers and has many consequences. I will endeavour to thrill you a little by giving a few such consequences after we have proved the theorem.

The Fundamental Theorem of Arithmetic

Without further ado then, let us state and prove the theorem.

THEOREM 11.1 (Fundamental Theorem of Arithmetic)
Let n be an integer with $n \geq 2$.
   (I) *Then n is equal to a product of prime numbers: we have*

$$n = p_1 \ldots p_k$$

where $p_1, \ldots, p_k$ are primes and $p_1 \leq p_2 \leq \ldots \leq p_k$.
   (II) *This prime factorization of n is unique: in other words, if*

$$n = p_1 \ldots p_k = q_1 \ldots q_l$$

where the p_is and q_is are all prime, $p_1 \leq p_2 \leq \ldots \leq p_k$ and $q_1 \leq q_2 \leq \ldots \leq q_l$, then

$$k = l \quad and \quad p_i = q_i \quad for\ all\ i = 1, \ldots, k.$$

The point about specifying that $p_1 \leq p_2 \leq \ldots \leq p_k$ is that this condition determines the order in which we write down the primes in the factorization of n. For example, 28 can be written as a product of primes in several ways: $2 \times 7 \times 2, 7 \times 2 \times 2$ and $2 \times 2 \times 7$. But if we specify that the prime factors have to increase or stay the same, then the only factorization is $28 = 2 \times 2 \times 7$.

PROOF   Part (I) is just Proposition 8.1.

Now for the uniqueness part (II). We prove this by contradiction. So suppose there is some integer n which has two different prime factorizations, say

$$n = p_1 \ldots p_k = q_1 \ldots q_l$$

where $p_1 \leq p_2 \leq \ldots \leq p_k$, $q_1 \leq q_2 \leq \ldots \leq q_l$, and the list of primes $p_1, \ldots, p_k$ is not the same list as $q_1, \ldots, q_l$.

Now in the equation $p_1 \ldots p_k = q_1 \ldots q_l$, cancel any primes that are common to both sides. Since we are assuming the two factorizations are different, not all the primes cancel, and we end up with an equation

$$r_1 \ldots r_a = s_1 \ldots s_b,$$

where each $r_i \in \{p_1, \ldots, p_k\}$, each $s_i \in \{q_1, \ldots, q_l\}$, and none of the r_is is equal to any of the s_is (i.e., $r_i \neq s_j$ for all i, j).

Now we obtain a contradiction. Certainly r_1 divides $r_1 \ldots r_a$, hence r_1 divides $s_1 \ldots s_b$. By Proposition 10.6, this implies that $r_1 | s_j$ for some j. However, s_j is prime, so its only divisors are 1 and s_j, and hence $r_1 = s_j$. But we know that none of the r_is is equal to any of the s_is, so this is a contradiction. This completes the proof of (II).   ∎

Of course, in the prime factorization given in part (I) of Theorem 11.1, some of the p_is may be equal to each other. If we collect these up, we obtain a unique prime factorization of the form

$$n = p_1^{a_1} p_2^{a_2} \ldots p_m^{a_m},$$

where $p_1 < p_2 < \ldots < p_m$ and the a_is are positive integers.

Some Consequences of the Fundamental Theorem

First, here is an application of the Fundamental Theorem of Arithmetic that looks rather more obvious than it really is.

PROPOSITION 11.1

Let $n = p_1^{a_1} p_2^{a_2} \dots p_m^{a_m}$, where the p_is are prime, $p_1 < p_2 < \dots < p_m$ and the a_is are positive integers. If m divides n, then

$$m = p_1^{b_1} p_2^{b_2} \dots p_m^{b_m} \quad \text{with} \quad 0 \le b_i \le a_i \text{ for all } i.$$

For example, the only divisors of $2^{100} 3^2$ are the numbers $2^a 3^b$, where $0 \le a \le 100, 0 \le b \le 2$.

PROOF If $m | n$, then $n = mc$ for some integer c. Let $m = q_1^{c_1} \dots q_k^{c_k}$, $c = r_1^{d_1} \dots r_l^{d_l}$ be the prime factorizations of m, c. Then $n = mc$ gives the equation

$$p_1^{a_1} p_2^{a_2} \dots p_m^{a_m} = q_1^{c_1} \dots q_k^{c_k} r_1^{d_1} \dots r_l^{d_l}.$$

By the Fundamental Theorem 11.1, the primes, and the powers to which they occur, must be identical on both sides. Hence, each q_i is equal to some p_j, and its power c_i is at most a_j. In other words, the conclusion of the proposition holds. ∎

We can use this to prove some further obvious-looking facts about integers. Define the *least common multiple* of two positive integers a and b, denoted by $lcm(a,b)$, to be the smallest positive integer that is divisible by both a and b. For example, $lcm(15,21) = 105$.

PROPOSITION 11.2

Let $a, b \ge 2$ be integers with prime factorizations

$$a = p_1^{r_1} p_2^{r_2} \dots p_m^{r_m}, \quad b = p_1^{s_1} p_2^{s_2} \dots p_m^{s_m}$$

where the p_i are distinct primes and all $r_i, s_i \ge 0$ (we allow some of the r_i and s_i to be 0). Then

(i) $hcf(a,b) = p_1^{\min(r_1, s_1)} \dots p_m^{\min(r_m, s_m)}$

(ii) $lcm(a,b) = p_1^{\max(r_1, s_1)} \dots p_m^{\max(r_m, s_m)}$

(iii) $lcm(a,b) = ab/hcf(a,b)$.

PROOF In part (i), the product on the right-hand side divides both a and b and is the largest such integer, by Proposition 11.1. And in part (ii), the product on the right-hand side is a multiple of both a and b and is the smallest such positive integer, again by Proposition 11.1. Finally, if we take the product of the right-hand sides in (i) and (ii), then we obtain

$$p_1^{\min(r_1,s_1)+\max(r_1,s_1)} \dots p_m^{\min(r_m,s_m)+\max(r_m,s_m)},$$

which is equal to ab since $\min(r_i,s_i)+\max(r_i,s_i) = r_i+s_i$. ∎

Here is our next application of the Fundamental Theorem of Arithmetic.

PROPOSITION 11.3
Let n be a positive integer. Then $\sqrt{n}$ is rational if and only if n is a perfect square (i.e., $n = m^2$ for some integer m).

PROOF The right-to-left implication is obvious: if $n = m^2$ with $m \in \mathbb{Z}$, then $\sqrt{n} = |m| \in \mathbb{Z}$ is certainly rational.

The left-to-right implication is much less clear. Suppose $\sqrt{n}$ is rational, say

$$\sqrt{n} = \frac{r}{s}$$

where $r,s \in \mathbb{Z}$. Squaring, we get $ns^2 = r^2$. Now consider prime factorizations. Each prime in the factorization of r^2 appears to an even power (since if $r = p_1^{a_1} \dots p_k^{a_k}$ then $r^2 = p_1^{2a_1} \dots p_k^{2a_k}$). The same holds for the primes in the factorization of s^2. Hence, by the Fundamental Theorem, each prime factor of n must also occur to an even power — say $n = q_1^{2b_1} \dots q_l^{2b_l}$. Then $n = m^2$, where $m = q_1^{b_1} \dots q_l^{b_l} \in \mathbb{Z}$. ∎

A similar argument applies to the rationality of cube roots, and more generally, n^{th} roots (see Exercise 5 at the end of the chapter).

Now for our final consequence of the Fundamental Theorem 11.1. Again it looks rather innocent, but in the example following the proposition we shall give a striking application of it.

In the statement, when we say a positive integer is a square (or an nth power), we mean that it is the square of an integer (or the nth power of an integer).

PROPOSITION 11.4
Let a and b be positive integers that are coprime to each other.

(a) If ab is a square, then both a and b are also squares.

(b) More generally, if ab is an nth power (for some positive integer n), then both a and b are also n^{th} powers.

PROOF (a) Let the prime factorizations of a, b be

$$a = p_1^{d_1} \dots p_k^{d_k}, \quad b = q_1^{e_1} \dots q_l^{e_l}$$

(where $p_1 < \ldots < p_k$ and $q_1 < \ldots < q_l$). If ab is a square, then $ab = c^2$ for some integer c; let c have prime factorization $c = r_1^{f_1} \ldots r_m^{f_m}$. Then $ab = c^2$ gives the equation

$$p_1^{d_1} \ldots p_k^{d_k} q_1^{e_1} \ldots q_l^{e_l} = r_1^{2f_1} \ldots r_m^{2f_m}.$$

Since a and b are coprime to each other, none of the p_is are equal to any of the q_is. Hence, the Fundamental Theorem 11.1 implies that each p_i is equal to some r_j, and the corresponding powers d_i and $2f_j$ are equal; and likewise for the q_is and their powers.

We conclude that all the powers d_i, e_i are even numbers — say $d_i = 2d_i', e_i = 2e_i'$. This means that

$$a = \left(p_1^{d_1'} \ldots p_k^{d_k'} \right)^2, \quad b = \left(q_1^{e_1'} \ldots q_l^{e_l'} \right)^2,$$

so a and b are squares.

(b) The argument for (b) is the same as for (a): an equation $ab = c^n$ gives an equality

$$p_1^{d_1} \ldots p_k^{d_k} q_1^{e_1} \ldots q_l^{e_l} = r_1^{nf_1} \ldots r_m^{nf_m}.$$

The Fundamental Theorem implies that each power d_i, e_i is a multiple of n, and hence a, b are both n^{th} powers.   ∎

Example 11.1

Here is an innocent little question about the integers:

Can a non-zero even square exceed a cube by 1?

(The non-zero even squares are of course the integers $4, 16, 64, 100, 144, \ldots$ and the cubes are $\ldots, -8, -1, 0, 1, 8, 27, \ldots$.)

In other words, we are asking whether the equation

$$4x^2 = y^3 + 1 \tag{11.1}$$

has any solutions with x, y both non-zero integers. This is an example of a *Diophantine equation*. In general, a Diophantine equation is an equation for which the solutions are required to be integers. Most Diophantine equations are very hard, or impossible, to solve — for instance, even the equation $x^2 = y^3 + k$ has not been completely solved for all values of k. However, I have chosen a nice example, in that Equation (11.1) can be solved fairly easily (as you will see), but the solution is not totally trivial and involves use of the consequence 11.4 of the Fundamental Theorem 11.1.

Let us then go about solving Equation (11.1) for $x, y \in \mathbb{Z}$. First we rewrite it as $y^3 = 4x^2 - 1$ and then cleverly factorize the right-hand side to get

$$y^3 = (2x+1)(2x-1).$$

The factors $2x+1, 2x-1$ are both odd integers, and their highest common factor divides their difference, which is 2. Hence

$$\mathrm{hcf}(2x+1, 2x-1) = 1.$$

Thus, $2x+1$ and $2x-1$ are coprime to each other, and their product is y^3, a cube. By Proposition 11.4(b), it follows that $2x+1$ and $2x-1$ are themselves both cubes. However, from the list of cubes $\ldots, -8, -1, 0, 1, 8, 27, \ldots$ it is apparent that the only two cubes that differ by 2 are $1, -1$. Therefore, $x = 0$ and we have shown that the only even square that exceeds a cube by 1 is 0. In other words, there are no non-zero such squares.

Exercises for Chapter 11

1. Find the prime factorization of 111111.

2. (a) Which positive integers have exactly three positive divisors?

   (b) Which positive integers have exactly four positive divisors?

3. Suppose $n \geq 2$ is an integer with the property that whenever a prime p divides n, p^2 also divides n (i.e., all primes in the prime factorization of n appear at least to the power 2). Prove that n can be written as the product of a square and a cube.

4. Prove that $lcm(a,b) = ab/hcf(a,b)$ for any positive integers a, b without using prime factorization.

5. (a) Prove that $2^{\frac{1}{3}}$ and $3^{\frac{1}{3}}$ are irrational.

   (b) Let m and n be positive integers. Prove that $m^{\frac{1}{n}}$ is rational if and only if m is an nth power (i.e., $m = c^n$ for some integer c).

6. Let E be the set of all positive even integers. We call a number e in E "prima" if e cannot be expressed as a product of two other members of E.

   (i) Show that 6 is prima but 4 is not.

(ii) What is the general form of a prima in E?

(iii) Prove that every element of E is equal to a product of primas.

(iv) Give an example to show that E does not satisfy a "unique prima factorization theorem" (i.e., find an element of E that has two different factorizations as a product of primas).

7. (a) Which pairs of positive integers m, n have $hcf(m,n) = 50$ and $lcm(m,n) = 1500$?

(b) Show that if m, n are positive integers, then $hcf(m,n)$ divides $lcm(m,n)$. When does $hcf(m,n) = lcm(m,n)$?

(c) Show that if m, n are positive integers, then there are coprime integers x, y such that x divides m, y divides n, and $xy = lcm(m,n)$.

8. Find all solutions $x, y \in \mathbb{Z}$ to the following Diophantine equations:

(a) $x^2 = y^3$.

(b) $x^2 - x = y^3$.

(c) $x^2 = y^4 - 77$.

(d) $x^3 = 4y^2 + 4y - 3$.

9. Languishing in his prison cell, critic Ivor Smallbrain is dreaming. In his dream he is on the Pacific island of Nefertiti, eating coconuts on a beach by a calm blue lagoon. Suddenly the king of Nefertiti approaches him, saying, "Your head will be chopped off unless you answer this riddle: Is it possible for the sixth power of an integer to exceed the fifth power of another integer by 16?" Feverishly, Ivor writes some calculations in the sand and eventually answers, "Oh, Great King, no it is not possible." The king rejoinders, "You are correct, but you will be beheaded anyway." The executioner's axe is just coming down when Ivor wakes up. He wonders whether his answer to the king was really correct.

Prove that Ivor was indeed correct.

Chapter 12

More on Prime Numbers

As you are probably beginning to appreciate, the prime numbers are fundamental to our understanding of the integers. In this chapter we will discuss a few basic results concerning the primes, and also hint at the vast array of questions, some solved, some unsolved, in current research into prime numbers.

The first few primes are

$$2, 3, 5, 7, 11, 13, 17, 19, 23, 29, 31, 37, 41, 43, \ldots$$

It is quite simple to carry on a long way with this list, particularly if you have a computer at hand. How would you do this? The easiest way is probably to test each successive integer n for primality, by checking, for each prime $p \leq \sqrt{n}$, whether p divides n (such primes p will of course already be in your list). If none of these primes p divides n, then n is prime — see Exercise 2 at the end of the chapter. Some more sophisticated methods for primality testing will be discussed at the end of Chapter 14.

Probably the first and most basic question to ask is: Does this list ever stop? In other words, is there a *largest* prime number, or does the list of primes go on forever? The answer is provided by the following famous theorem of Euclid (300 BC).

THEOREM 12.1
There are infinitely many prime numbers.

PROOF   This is one of the classic proofs by contradiction. Assume the result is false — that is, there are only finitely many primes. This means that we can make a finite list

$$p_1, p_2, p_3, \ldots, p_n$$

of *all* the prime numbers. Now define a positive integer

$$N = p_1 p_2 p_3 \ldots p_n + 1.$$

By Proposition 8.1, N is equal to a product of primes, say $N = q_1 \ldots q_r$ with all q_i prime. As q_1 is prime, it belongs to the above list of all primes, so $q_1 = p_i$ for some i.

Now q_1 divides N, and hence p_i divides N. Also p_i divides $p_1 p_2 \ldots p_n$, which is equal to $N - 1$. Thus, p_i divides both N and $N - 1$. But this implies that p_i divides the difference between these numbers, namely 1. This is a contradiction. ∎

Theorem 12.1 is of course not the end of the story about the primes — it is really the beginning. A natural question to ask that flows from the theorem is: what *proportion* of all positive integers are prime? On the face of it this question makes no sense, as the integers and the primes are both infinite sets. But one can make a sensible question by asking

Given a positive integer n, how many of the numbers $1, 2, 3, \ldots, n$ *are prime?*

Is there any reason to expect to be able to answer this question? On the face of it, no. If you stare at a long list of primes, you will see that the sequence is very irregular, and it is very difficult to see any pattern at all in it. (See, for example, Exercise 6 at the end of the chapter.) Why on earth should there then be a nice formula for the number of primes up to n?

The amazing thing is that there *is* such a formula, albeit an "asymptotic" one. (I will explain this word later.) The great Gauss, by calculating a lot with lists of primes (and also by having a lot of brilliant thoughts), formed the incredible conjecture (i.e., informed guess) that the number of primes up to n should be pretty close to the formula

$$\frac{n}{\log_e n}.$$

To understand this a little, compare the number of primes up to 10^6 (namely, 78498) with the value of $\frac{10^6}{\log 10^6}$ (namely, 72382.4). The *difference* between these two numbers, about 6000, appears to be quite large; but their *ratio* is 1.085, quite close to 1. It was on the ratio, rather than the difference, that Gauss concentrated his mind: his conjecture was that the ratio of the number of primes up to n and the expression $\frac{n}{\log_e n}$ should get closer and closer to 1 as n gets larger and larger. (Formally, this ratio *tends to* 1 *as n tends to infinity*.)

Gauss did not actually manage to prove his conjecture. The world had to wait until 1896, when a Frenchman, Hadamard, and a Belgian, de la Vallée-Poussin, both produced proofs of what is now known as the Prime Number Theorem:

THEOREM 12.2

For a positive integer n, let $\pi(n)$ *be the number of primes up to n. Then the ratio of* $\pi(n)$ *and* $\frac{n}{\log_e n}$ *tends to 1 as n tends to infinity (i.e., the ratio can be made as close as we like to 1 provided n is large enough).*

The proof of this result uses some quite sophisticated tools of analysis. Nevertheless, if you are lucky you might get the chance to see a proof in an undergraduate course later in your studies — in other words, it is not *that* difficult!

You should not think that every question about the primes can be answered (if not by you, then by some expert or other). On the contrary, many basic questions about the primes are unsolved to this day, despite being studied for many years. Let me finish this chapter by mentioning a couple of the most famous such problems.

The Goldbach conjecture If you do some calculations, or program your computer, you will find that any reasonably small even positive integer greater than 2 can be expressed as a sum of two primes. For example,

$$10 = 7 + 3, \ 50 = 43 + 7, \ 100 = 97 + 3, \ 8000 = 3943 + 4057$$

and so on. Based on this evidence, it seems reasonable to conjecture that *every* even positive integer is the sum of two primes. This is the Goldbach conjecture, and it is unsolved to this day.

The twin prime conjecture If p and $p + 2$ are both prime numbers, we call them *twin primes*. For example, here are some twin primes:

$$3,5; \ 5,7; \ 11,13; \ 71,73; \ 1997,1999.$$

If you stare at a list of prime numbers, you will find many pairs of twin primes, getting larger and larger. One feels that there should be infinitely many twin primes, and indeed, that statement is known as the twin prime conjecture. Can one prove the twin prime conjecture using a proof like Euclid's in Theorem 12.1? Unfortunately not — indeed, no one has come up with any sort of proof, and the conjecture remains unsolved to this day.

Exercises for Chapter 12

1. Prove Liebeck's *triplet prime conjecture*: the only triplet of primes of the form $p, p+2, p+4$ is $\{3,5,7\}$.

2. Let n be an integer with $n \geq 2$. Suppose that for every prime $p \leq \sqrt{n}$, p does not divide n. Prove that n is prime.

   Is 221 prime? Is 223 prime?

3. For a positive integer n, define $\phi(n)$ to be the number of positive integers $a < n$ such that $\mathrm{hcf}(a,n) = 1$. (For example, $\phi(2) = 1$, $\phi(3) = 2$, $\phi(4) = 2$.)

   Work out $\phi(n)$ for $n = 5, 6, \ldots, 10$.

   If p is a prime, show that $\phi(p) = p - 1$ and, more generally, that $\phi(p^r) = p^r - p^{r-1}$.

4. Use the idea of the proof of Euclid's Theorem 12.1 to prove that there are infinitely many primes of the form $4k + 3$ (where k is an integer).

5. There has been quite a bit of work over the years on trying to find a nice formula that takes many prime values. For example, $x^2 + x + 41$ is prime for all integers x such that $-40 \leq x < 40$. (You may like to check this!) However:

   Find an integer x coprime to 41 such that $x^2 + x + 41$ is not prime.

6. On his release from prison, critic Ivor Smallbrain rushes out to see the latest film, *Prime and Prejudice*. During the film Ivor attempts to think of ten consecutive positive integers, none of which is prime. He fails.

   Help Ivor by showing that if $N = 11! + 2$, then none of the numbers $N, N+1, N+2, \ldots, N+9$ is prime.

   More generally, show that for any $n \in \mathbb{N}$ there is a sequence of n consecutive positive integers, none of which is prime. (Hence, there are arbitrarily large "gaps" in the sequence of primes.)

Chapter *13*

Congruence of Integers

In this chapter we introduce another method for studying the integers, called congruence. Let us go straight into the definition.

DEFINITION   Let m be a positive integer. For $a, b \in \mathbb{Z}$, if m divides $b - a$ we write $a \equiv b$ mod m and say a is congruent to b modulo m.

For example,

$$5 \equiv 1 \bmod 2, \quad 12 \equiv 17 \bmod 5, \quad 91 \equiv -17 \bmod 12, \quad 531 \not\equiv 0 \bmod 4.$$

PROPOSITION 13.1
Every integer is congruent to exactly one of the numbers $0, 1, 2, \ldots, m - 1$ modulo m.

PROOF   Let $x \in \mathbb{Z}$. By Proposition 10.1, there are integers q, r such that

$$x = qm + r \quad \text{with} \quad 0 \leq r < m.$$

Then $x - r = qm$, so m divides $x - r$, and hence by the above definition, $x \equiv r$ mod m. Since r is one of the numbers $0, 1, 2, \ldots, m - 1$, the proposition follows.   ∎

Example 13.1
(1) Every integer is congruent to 0 or 1 modulo 2. Indeed, all even integers are congruent to 0 modulo 2 and all odd integers to 1 modulo 2.

(2) Every integer is congruent to 0, 1, 2 or 3 modulo 4. More specifically, every even integer is congruent to 0 or 2 modulo 4 and every odd integer to 1 or 3 modulo 4.

(3) My clock is now showing the time as 2:00 a.m. What time will it be showing in 4803 hours? Since $4803 \equiv 3$ mod 24, it will be showing a

time 3 hours later than the current time, namely 5:00 a.m. (But I hope I will not be awake to see it.)

The next result will be quite useful for our later work involving manipulation of congruences.

PROPOSITION 13.2

Let m be a positive integer. The following are true, for all $a, b, c \in \mathbb{Z}$:

(1) $a \equiv a \bmod m$,

(2) *if* $a \equiv b \bmod m$ *then* $b \equiv a \bmod m$,

(3) *if* $a \equiv b \bmod m$ *and* $b \equiv c \bmod m$, *then* $a \equiv c \bmod m$.

PROOF   (1) Since $m|0$ we have $m|a-a$, and hence $a \equiv a \bmod m$.

(2) If $a \equiv b \bmod m$ then $m|b-a$, so $m|a-b$, and hence $b \equiv a \bmod m$.

(3) If $a \equiv b \bmod m$ and $b \equiv c \bmod m$, then $m|b-a$ and $m|c-b$; say $b-a = km,\, c-b = lm$. Then $c-a = (k+l)m$, so $m|c-a$, and hence $a \equiv c \bmod m$. ∎

Arithmetic with Congruences

Congruence is a notation that conveniently records various divisibility properties of integers. This notation comes into its own when we do arithmetic with congruences, as we show is possible in the next two results. The first shows that congruences modulo m can be added and multiplied.

PROPOSITION 13.3

Suppose $a \equiv b \bmod m$ and $c \equiv d \bmod m$. Then

$$a + c \equiv b + d \bmod m \quad \text{and} \quad ac \equiv bd \bmod m.$$

PROOF   We are given that $m|b-a$ and $m|d-c$. Say $b-a = km$ and $d-c = lm$, where $k, l \in \mathbb{Z}$. Then

$$(b+d) - (a+c) = (k+l)m$$

and hence $a + c \equiv b + d \bmod m$. And

$$bd - ac = (a+km)(c+lm) - ac = m(al + ck + klm),$$

which implies that $ac \equiv bd$ mod m. ∎

PROPOSITION 13.4
If $a \equiv b$ mod m, and n is a positive integer, then

$$a^n \equiv b^n \text{ mod } m.$$

PROOF We prove this by induction. Let $P(n)$ be the statement of the proposition. Then $P(1)$ is obviously true.

Now suppose $P(n)$ is true, so $a^n \equiv b^n$ mod m. As $a \equiv b$ mod m, we can use Proposition 13.3 to multiply these congruences and get $a^{n+1} \equiv b^{n+1}$ mod m, which is $P(n+1)$. Hence, $P(n)$ is true for all n by induction. ∎

These results give us some powerful methods for using congruences, as we shall now attempt to demonstrate with a few examples.

Example 13.2
Find the remainder r (between 0 and 6) that we get when we divide 6^{82} by 7.

Answer We start with the congruence $6 \equiv -1$ mod 7. By Proposition 13.4, we can raise this to the power 82, to get $6^{82} \equiv (-1)^{82}$ mod 7, hence $6^{82} \equiv 1$ mod 7. This means that 7 divides $6^{82} - 1$; hence $6^{82} = 7q + 1$ for some $q \in \mathbb{Z}$, and so the remainder is 1.

Example 13.3
Find the remainder r (between 0 and 12) that we get when we divide 6^{82} by 13.

Answer This is not quite so easy as the previous example. We employ a general method, which involves "successive squaring" of the congruence $6 \equiv 6$ mod 13. Squaring once, we get $6^2 \equiv 36$ mod 13; since $36 \equiv -3$ mod 13, Proposition 13.2(3) gives $6^2 \equiv -3$ mod 13. Successive squaring like this yields:

$$6^2 \equiv -3 \text{ mod } 13,$$
$$6^4 \equiv 9 \text{ mod } 13,$$
$$6^8 \equiv 3 \text{ mod } 13,$$
$$6^{16} \equiv 9 \text{ mod } 13,$$
$$6^{32} \equiv 3 \text{ mod } 13,$$

$$6^{64} \equiv 9 \text{ mod } 13.$$

Now $6^{82} = 6^{64}6^{16}6^2$. Multiplying the above congruences for $6^{64}, 6^{16}$ and 6^2, we get

$$6^{82} \equiv 9 \cdot 9 \cdot (-3) \text{ mod } 13.$$

Now $9 \cdot (-3) = -27 \equiv -1 \text{ mod } 13$, so $6^{82} \equiv -9 \equiv 4 \text{ mod } 13$. Hence, the required remainder is 4.

The method given in this example is called the *method of successive squares* and always works to yield the congruence of a large power, given some effort. To work out the congruence modulo m of a power x^k of some integer x, express k as a sum of powers of 2 (you might have met this before as "writing k in base 2" or some such phrase) — say $k = 2^{a_1} + \cdots + 2^{a_r}$; then successively square to work out the congruences modulo m of the powers $x^{2^{a_1}}, \ldots, x^{2^{a_r}}$, and multiply these together to obtain the answer.

Sometimes this effort can be reduced with some clever trickery, as in Example 13.2.

Example 13.4
Show that no integer square is congruent to 2 modulo 3. (In other words, the sequence $2, 5, 8, 11, 14, 17, \ldots$ contains no squares.)

Answer Consider an integer square n^2 (where $n \in \mathbb{Z}$). By Proposition 13.1, n is congruent to 0, 1 or 2 modulo 3. If $n \equiv 0 \text{ mod } 3$, then by Proposition 13.4, $n^2 \equiv 0 \text{ mod } 3$; if $n \equiv 1 \text{ mod } 3$, then $n^2 \equiv 1 \text{ mod } 3$; and if $n \equiv 2 \text{ mod } 3$, then $n^2 \equiv 4 \text{ mod } 3$, and hence [using 13.2(3)] $n^2 \equiv 1 \text{ mod } 3$. This shows that integer squares are congruent to 0 or 1 modulo 3.

Example 13.5
Show that every odd integer square is congruent to 1 modulo 4.

Answer This is similar to the previous example. Let n be an odd integer. Then n is congruent to 1 or 3 modulo 4, so n^2 is congruent to 1 or 9 modulo 4, hence to 1 modulo 4.

Example 13.6
The "rule of 3" You may have come across a simple rule for testing whether an integer is divisible by 3: add up its digits, and if the sum is divisible by 3 then the integer is divisible by 3. Here is a quick explanation of why this rule works.

Let n be an integer, with digits $a_r a_{r-1} \ldots a_0$, so

$$n = a_0 + 10a_1 + 10^2 a_2 + \cdots + 10^r a_r.$$

Now $10 \equiv 1 \bmod 3$; hence, by Proposition 13.4, $10^k \equiv 1 \bmod 3$ for any positive integer k. Multiplying this by the congruence $a_k \equiv a_k \bmod 3$ gives $10^k a_k \equiv a_k \bmod 3$. It follows that

$$n \equiv a_0 + a_1 + \cdots + a_r \bmod 3.$$

Hence, $n \equiv 0 \bmod 3$ if and only if the sum of its digits $a_0 + \cdots + a_r \equiv 0 \bmod 3$. This is the "rule of 3."

The same method proves the "rule of 9": an integer is divisible by 9 if and only if the sum of its digits is divisible by 9. There is also a "rule of 11," which is not quite so obvious: an integer n with digits $a_r \ldots a_1 a_0$ is divisible by 11 if and only if the expression $a_0 - a_1 + a_2 - \cdots + (-1)^r a_r$ is divisible by 11. Proving this is Exercise 4 at the end of the chapter. You will also find other rules in Exercise 5.

Unlike adding and multiplying, we can't always divide congruences modulo m. For example, $10 \equiv 6 \bmod 4$, but we can't divide this by 2 to deduce that $5 \equiv 3 \bmod 4$. However, the next result shows that there are some circumstances in which we can divide congruences.

PROPOSITION 13.5

(1) Let a and m be coprime integers. If $x, y \in \mathbb{Z}$ are such that $xa \equiv ya \bmod m$, then $x \equiv y \bmod m$.

(2) Let p be a prime, and let a be an integer that is not divisible by p. If $x, y \in \mathbb{Z}$ are such that $xa \equiv ya \bmod p$, then $x \equiv y \bmod p$.

PROOF (1) Assume that $xa \equiv ya \bmod m$. Then m divides $xa - ya = (x - y)a$. Since a, m are coprime, Proposition 10.5(a) implies that m divides $x - y$. In other words, $x \equiv y \bmod m$.

Part (2) is immediate from part (1), since if a is not divisible by a prime p, then a and p are coprime. ∎

Congruence Equations

Let m be a positive integer and let $a, b \in \mathbb{Z}$. Consider the equation

$$ax \equiv b \bmod m$$

to be solved for $x \in \mathbb{Z}$. Such an equation is called a linear congruence equation. When does such an equation have a solution?

Example 13.7

(1) Consider the congruence equation

$$4x \equiv 2 \bmod 28.$$

If $x \in \mathbb{Z}$ is a solution to this, then $4x = 2 + 28n$ for some integer n, which is impossible since the left-hand side is divisible by 4, whereas the right-hand side is not. So this congruence equation has no solutions.

(2) Now consider the equation

$$13x \equiv 2 \bmod 31.$$

We shall show that this equation has a solution. Observe that $\mathrm{hcf}(13, 31) = 1$; hence, by Proposition 10.3, there are integers s, t such that

$$1 = 13s + 31t.$$

Therefore, $13s = 1 - 31t$, which means that $13s \equiv 1 \bmod 31$. Multiplying this congruence by 2, we get

$$13 \cdot (2s) \equiv 2 \bmod 31.$$

In other words, $x = 2s$ is a solution to the original congruence equation.

Here is a general result telling us exactly when linear congruence equations have solutions.

PROPOSITION 13.6

The congruence equation

$$ax \equiv b \bmod m$$

has a solution $x \in \mathbb{Z}$ if and only if $\mathrm{hcf}\,(a, m)$ divides b.

PROOF   Write $d = \mathrm{hcf}(a, m)$. First let us prove the left-to-right implication. So suppose the equation has a solution $x \in \mathbb{Z}$. Then $ax = qm + b$ for some integer q. Since $d|a$ and $d|m$, it follows that $d|b$.

Now for the right-to-left implication. Suppose $d|b$, say $b = kd$. By Proposition 10.3, there are integers s, t such that $d = sa + tm$. Multiplying through by k gives $b = kd = k(sa + tm)$. Hence,

$$aks = b - ktm \equiv b \bmod m.$$

In other words, $x = ks$ is a solution to the congruence equation.   ∎

We shall see some different types of congruence equations in the next chapter.

The System $\mathbb{Z}_m$

The properties of congruence modulo m can be encapsulated rather neatly by defining a new system, denoted by $\mathbb{Z}_m$, which we can think of as "the integers modulo m." Before defining this in general, here is an example.

Example 13.8

Take $m = 4$. Let $\mathbb{Z}_4$ be the set consisting of the four symbols $\bar{0}, \bar{1}, \bar{2}, \bar{3}$, and define addition and multiplication of any two symbols in $\mathbb{Z}_4$ to be the same as adding or multiplying the two corresponding integers in $\mathbb{Z}$, except that we make sure the answer is again in $\mathbb{Z}_4$ by taking congruences modulo 4. For example, to add $\bar{2}$ and $\bar{3}$ in $\mathbb{Z}_4$, we first note that the sum of the correpsonding integers 2 and 3 in $\mathbb{Z}$ is 5; the number in $\mathbb{Z}_4$ that is congruent to 5 modulo 4 is 1, so in $\mathbb{Z}_4$ we define the sum $\bar{2} + \bar{3}$ to be $\bar{1}$. Here are some other examples of addition and multiplication in $\mathbb{Z}_4$:

$$\bar{1} + \bar{2} = \bar{3}, \ \bar{0} + \bar{3} = \bar{3}, \ \bar{2} + \bar{2} = \bar{0}, \ \bar{3} + \bar{3} = \bar{2},$$

and

$$\bar{0} \times \bar{3} = \bar{0}, \ \bar{2} \times \bar{3} = \bar{2}, \ \bar{3} \times \bar{3} = \bar{1}.$$

The full addition and multiplication tables for $\mathbb{Z}_4$ are as follows:

+	$\bar{0}$	$\bar{1}$	$\bar{2}$	$\bar{3}$
$\bar{0}$	$\bar{0}$	$\bar{1}$	$\bar{2}$	$\bar{3}$
$\bar{1}$	$\bar{1}$	$\bar{2}$	$\bar{3}$	$\bar{0}$
$\bar{2}$	$\bar{2}$	$\bar{3}$	$\bar{0}$	$\bar{1}$
$\bar{3}$	$\bar{3}$	$\bar{0}$	$\bar{1}$	$\bar{2}$

×	$\bar{0}$	$\bar{1}$	$\bar{2}$	$\bar{3}$
$\bar{0}$	$\bar{0}$	$\bar{0}$	$\bar{0}$	$\bar{0}$
$\bar{1}$	$\bar{0}$	$\bar{1}$	$\bar{2}$	$\bar{3}$
$\bar{2}$	$\bar{0}$	$\bar{2}$	$\bar{0}$	$\bar{2}$
$\bar{3}$	$\bar{0}$	$\bar{3}$	$\bar{2}$	$\bar{1}$

In general, for a fixed positive integer m, the system $\mathbb{Z}_m$ is defined in an entirely similar way. We take $\mathbb{Z}_m$ to be the set consisting of the m symbols $\bar{0}, \bar{1}, \bar{2}, \ldots, \overline{m-1}$, and define addition and multiplication of any two symbols in $\mathbb{Z}_m$ to be the same as for the corresponding integers in $\mathbb{Z}$, except that we make sure the answer is again in $\mathbb{Z}_m$ by taking congruences modulo m. More formally, for $\bar{x}, \bar{y} \in \mathbb{Z}_m$, the sum and product of $\bar{x}$ and $\bar{y}$ in $\mathbb{Z}_m$ are defined to be the symbols $\bar{k}, \bar{l} \in \mathbb{Z}_m$ such that $x + y \equiv k \bmod m$ and $xy \equiv l \bmod m$, respectively. We'll write $\bar{x} + \bar{y} = \bar{k}$ and $\bar{x}\bar{y} = \bar{l}$ in $\mathbb{Z}_m$.

For example, in $\mathbb{Z}_5$ we have $\bar{4} + \bar{4} = \bar{3}$; in $\mathbb{Z}_7$ we have $\bar{5} \times \bar{3} = \bar{1}$; and in any $\mathbb{Z}_m$, we have $\overline{m-1} + \overline{m-2} = \overline{m-3}$ and $\overline{m-1}\,\overline{m-2} = \bar{2}$.

The system $\mathbb{Z}_m$ is quite a useful one. We can do quite a bit of algebra in it, such as taking powers and solving equations. Here are a few more examples.

Example 13.9

(1) We can define powers of elements of $\mathbb{Z}_m$ in a natural way: for $\bar{x} \in \mathbb{Z}_m$ and $r \in \mathbb{N}$, the rth power of $\bar{x}$ in $\mathbb{Z}_m$ is defined to be the symbol $\bar{k} \in \mathbb{Z}_m$ such that $x^r \equiv k \bmod m$. For example, in $\mathbb{Z}_5$, we have

$$\bar{2}^2 = \bar{4}, \ \bar{2}^3 = \bar{3}, \ \bar{2}^4 = \bar{1}.$$

Examples 13.2 and 13.3 show that in $\mathbb{Z}_7$, we have $\bar{6}^{82} = \bar{1}$, while in $\mathbb{Z}_{13}$, $\bar{6}^{82} = \bar{4}$.

(2) Proposition 13.6 tells us that for $\bar{a}, \bar{b} \in \mathbb{Z}_m$, the equation $\bar{a}\bar{x} = \bar{b}$ has a solution for $\bar{x} \in \mathbb{Z}_m$ if and only if $\text{hcf}(a,m)$ divides b. So, for example, the equation $\bar{2}\bar{x} = \bar{7}$ has a solution in $\mathbb{Z}_9$ (namely $\bar{x} = \bar{8}$), but not in $\mathbb{Z}_{10}$.

(3) Examples 13.4 and 13.5 show that the quadratic equation $\bar{x}^2 = \bar{2}$ has no solution in $\mathbb{Z}_3$, and $\bar{x}^2 = \bar{3}$ has no solution in $\mathbb{Z}_4$. In general, to see whether an equation in a variable $\bar{x}$ has a solution in $\mathbb{Z}_m$, unless we can think of anything better we can always just try substituting each of the m possible values for $\bar{x}$ and seeing if they work. For example, does the equation

$$\bar{x}^2 + \bar{3}\bar{x} + \bar{4} = \bar{0}$$

have a solution in $\mathbb{Z}_5$? Well, if we substitute the five possible values for $\bar{x}$ into the expression $\bar{x}^2 + \bar{3}\bar{x} + \bar{4}$ and work out the answer in $\mathbb{Z}_5$, here is what we get:

$\bar{x}$	$\bar{0}$	$\bar{1}$	$\bar{2}$	$\bar{3}$	$\bar{4}$
$\bar{x}^2 + \bar{3}\bar{x} + \bar{4}$	$\bar{4}$	$\bar{3}$	$\bar{4}$	$\bar{2}$	$\bar{2}$

So the answer is no, there is no solution in $\mathbb{Z}_5$. However, this equation does have solutions in $\mathbb{Z}_{11}$, for example — namely $\bar{x} = \bar{3}$ or $\bar{5}$.

Exercises for Chapter 13

1. (a) Find r with $0 \le r \le 10$ such that $7^{137} \equiv r \bmod 11$.

   (b) Find r with $0 \le r < 645$ such that $2^{81} \equiv r \bmod 645$.

   (c) Find the last two digits of 3^{124} (when expressed in decimal notation).

   (d) Show that there is a multiple of 21 which has 241 as its last three digits.

2. Let p be a prime number and k a positive integer.

   (a) Show that if x is an integer such that $x^2 \equiv x \bmod p$, then $x \equiv 0$ or $1 \bmod p$.

   (b) Show that if x is an integer such that $x^2 \equiv x \bmod p^k$, then $x \equiv 0$ or $1 \bmod p^k$.

3. For each of the following congruence equations, either find a solution $x \in \mathbb{Z}$ or show that no solution exists:

(a) $99x \equiv 18 \bmod 30$.

(b) $91x \equiv 84 \bmod 143$.

(c) $x^2 \equiv 2 \bmod 5$.

(d) $x^2 + x + 1 \equiv 0 \bmod 5$.

(e) $x^2 + x + 1 \equiv 0 \bmod 7$.

4. (a) Prove the "rule of 9": an integer is divisible by 9 if and only if the sum of its digits is divisible by 9.

(b) Prove the "rule of 11" stated in Example 13.6. Use this rule to decide in your head whether the number 82918073579 is divisible by 11.

5. (a) Use the fact that 7 divides 1001 to find your own "rule of 7." Use your rule to work out the remainder when 6005004003002001 is divided by 7.

(b) 13 also divides 1001. Use this to get a rule of 13 and find the remainder when 6005004003002001 is divided by 13.

(c) Use the observation that $27 \times 37 = 999$ to work out a rule of 37, and find the remainder when 6005004003002001 is divided by 37.

6. Let p be a prime number, and let a be an integer that is not divisible by p. Prove that the congruence equation $ax \equiv 1 \bmod p$ has a solution $x \in \mathbb{Z}$.

7. Show that every square is congruent to 0, 1 or -1 modulo 5, and is congruent to 0, 1 or 4 modulo 8.

Suppose n is a positive integer such that both $2n + 1$ and $3n + 1$ are squares. Prove that n is divisible by 40.

Find a value of n such that $2n + 1$ and $3n + 1$ are squares. Can you find another value? (Calculators allowed!)

8. Find $\bar{x}, \bar{y} \in \mathbb{Z}_{15}$ such that $\bar{x}\bar{y} = \bar{0}$ but $\bar{x} \neq \bar{0}, \bar{y} \neq \bar{0}$.

Find a condition on m such that the equality $\bar{x}\bar{y} = \bar{0}$ in $\mathbb{Z}_m$ implies that either $\bar{x} = \bar{0}$ or $\bar{y} = \bar{0}$.

9. Let p be a prime and let $\bar{a}, \bar{b} \in \mathbb{Z}_p$, with $\bar{a} \neq \bar{0}$ and $\bar{b} \neq \bar{0}$. Prove that the equation $\bar{a}\bar{x} = \bar{b}$ has a solution for $\bar{x} \in \mathbb{Z}_p$.

10. Construct the addition and multiplication tables for $\mathbb{Z}_6$. Find all solutions in $\mathbb{Z}_6$ of the equation $\bar{x}^2 + \bar{x} = 0$.

11. It is Friday, May 6, 2005. Ivor Smallbrain is watching the famous movie *From Here to Infinity*. He is bored, and idly wonders what day of the week it will be on the same date in 1000 years' time (i.e., on May 6, 3005). He decides it will again be a Friday.

Is Ivor right? And what has this question got to do with congruence?

Chapter 14

More on Congruence

In this chapter we are going to see some further results about congruence of integers. Most of these are to do with working out the congruence of large powers of an integer modulo some given integer m. I showed you some ways of tackling this kind of question in the last chapter (see Examples 13.2 and 13.3). The first result of this chapter — Fermat's Little Theorem — is a general fact that makes powers rather easy to calculate when m is a prime number. The rest of the chapter consists mainly of applications of this theorem to solving some special types of congruence equations modulo a prime or a product of two primes, and also to the problem of finding large prime numbers using a computer. We'll make heavy use of all this material in the next chapter on secret codes.

Fermat's Little Theorem

This very nifty result was first found by the French mathematician Fermat around 1640. It is called Fermat's Little Theorem to distinguish it from the rather famous "Fermat's Last Theorem," which is somewhat harder to prove (although why the adjective "little" was chosen, rather than "large" or "medium-sized" or "nifty," is not clear to me).

Anyway, here it is.

THEOREM 14.1 (Fermat's Little Theorem)
Let p be a prime number, and let a be an integer that is not divisible by p. Then
$$a^{p-1} \equiv 1 \ mod \ p.$$

For example, applying the theorem with $p = 17$ tells us that $2^{16} \equiv 1$ mod 17, that $93^{16} \equiv 1$ mod 17, and indeed that $72307892^{16} \equiv 1$ mod 17. This makes congruences of powers modulo 17 fairly painless to calculate. For instance, let's work out 3^{972} modulo 17. Well, we know that $3^{16} \equiv 1 \bmod 17$. Dividing 16 into 972, we get $972 = 16 \cdot 60 + 12$, so

$$3^{972} = (3^{16})^{60} \cdot 3^{12} \equiv (1^{60}) \cdot 3^{12} \bmod 17.$$

So we only need to work out 3^{12} modulo 17. This is easily done using the method of successive squares explained in Example 13.3: we get $3^2 \equiv 9 \bmod 17$, so $3^4 \equiv 9^2 \equiv 13 \bmod 17$, so $3^8 \equiv 13^2 \equiv (-4)^2 \equiv 16 \bmod 17$. Hence

$$3^{972} \equiv 3^{12} \equiv 3^4 \cdot 3^8 \equiv 13 \cdot 16 \equiv 4 \bmod 17.$$

PROOF   Here's a proof of Fermat's Little Theorem. For an integer x, we shall write $x \pmod p$ to denote the integer k between 0 and $p-1$ such that $x \equiv k \bmod p$. (So, for example, $17 \pmod 7) = 3$.) Also, for integers $x_1, \ldots, x_k$, we write $x_1, \ldots, x_k \pmod p$ as an abbreviation for the list $x_1 \pmod p, \ldots, x_k \pmod p$. (So for example, the list $17, 32, -1 \pmod 7$ is $3, 4, 6$.)

Now let a, p be as in the theorem. Multiply each of the numbers $1, 2, 3, \ldots, p-1$ by a and take congruences modulo p to get the list

$$a, 2a, 3a, \ldots, (p-1)a \pmod p. \tag{14.1}$$

(For example, if $p = 7$ and $a = 3$, this list is $3, 6, 2, 5, 1, 4$.)

We claim that the list (14.1) consists of the numbers $1, 2, 3, \ldots, p-1$ in some order. To see this, note first that since p does not divide a, none of the numbers in the list is 0. Next, suppose two of the numbers in the list are equal, say $xa \pmod p) = ya \pmod p$ (where $1 \le x, y \le p-1$). Saying that $xa \pmod p) = ya \pmod p$ is the same as saying that $xa \equiv ya \bmod p$. By Proposition 13.5(2), this implies that $x \equiv y \bmod p$, and since x and y are between 1 and $p-1$, this means that $x = y$. This shows that the numbers listed in (14.1) are all different. As there are $p-1$ of them, and none of them is 0, they must be the numbers $1, 2, 3, \ldots, p-1$ in some order.

Now let's multiply together all the numbers in the list (14.1). The result is $a^{p-1} \cdot (p-1)! \pmod p$. Since the numbers in the list are just $1, 2, 3, \ldots p-1$ in some order, this product is also equal to $(p-1)! \pmod p$. In other words, $a^{p-1} \cdot (p-1)! \pmod p) = (p-1)! \pmod p$, which means that

$$a^{p-1} \cdot (p-1)! \equiv (p-1)! \bmod p.$$

Now p does not divide $(p-1)!$ (since none of the factors $p-1, p-2, \ldots 3, 2$ is divisible by p). Hence Proposition 13.5(2) allows us to cancel $(p-1)!$ in the above congruence equation to deduce that $a^{p-1} \equiv 1 \bmod p$. This completes the proof.   ∎

Notice that Fermat's Little Theorem implies that if p is prime, then $a^p \equiv a \bmod p$ for all integers a, regardless of whether a is divisible by p (since if p divides a then obviously $a^p \equiv a \equiv 0 \bmod p$). For a different proof of Fermat's Little Theorem, see Exercise 9 at the end of Chapter 16.

It is possible to use Fermat's Little Theorem to show that a number N is not a prime without actually finding any factors of N. For example, suppose we are trying to test whether the number 943 is prime. One test is to raise some number, say 2, to the power 942 and see whether the answer is congruent to 1 modulo 943; if it's not, then 943 is not a prime by Fermat's Little Theorem. In fact, using the method of successive squares (see Example 13.3) we can calculate that

$$2^{942} \equiv 496 \bmod 943,$$

showing that indeed 943 is not a prime.

Of course, for such a small number as 943 it would be much quicker to test it for primality by simply looking for prime factors less than $\sqrt{943}$. However, for really large numbers the above test can be rather effective. Let's call it the *Fermat test*.

The Fermat test is by no means always successful in detecting the non-primality of a number. For example, you will find that $2^{340} \equiv 1 \bmod 341$, which gives you no information about whether or not 341 is prime. (In fact, it is not prime, since $341 = 11 \cdot 31$.) Indeed, there are some numbers N that are not prime yet satisfy $a^{N-1} \equiv 1 \bmod N$ for all integers a coprime to N, and for these numbers the Fermat test will never detect their non-primality. The smallest such number is 561 (see Exercise 3 at the end of the chapter).

We shall see in the next chapter that it is very important to be able to find very large primes using computers, so finding good primality tests is vital. There are some clever refinements of the Fermat test that work well in practice, and I'll discuss these later in this chapter.

Before moving on to the next section, we note the following result, which is an easy consequence of Fermat's Little Theorem. We'll need this also in the next chapter.

PROPOSITION 14.1
Let p and q be distinct prime numbers, and let a be an integer that is not divisible by p or by q. Then

$$a^{(p-1)(q-1)} \equiv 1 \ mod \ pq.$$

PROOF By Fermat's Little Theorem we know that $a^{p-1} \equiv 1 \bmod p$. Taking $(q-1)^{th}$ powers of both sides (which we can do by Proposition 13.4), it follows that $a^{(p-1)(q-1)} \equiv 1 \bmod p$. Similarly $a^{q-1} \equiv 1 \bmod q$, and hence also $a^{(q-1)(p-1)} \equiv 1 \bmod q$. Therefore, both p and q divide

$a^{(p-1)(q-1)} - 1$, so pq divides this number (since p and q both appear in its prime factorization). In other words, $a^{(p-1)(q-1)} \equiv 1 \bmod pq$, as required. ∎

Finding k^{th} Roots Modulo m

We now consider congruence equations of the form

$$x^k \equiv b \bmod m,$$

where m, b and k are given integers and we want to solve for $x \pmod m$. We can regard any such solution x as a kth root of b modulo m, so solving this equation is equivalent to finding k^{th} roots modulo m.

In fact we now have enough theory in place to be able to do this under certain assumptions when m is either a prime or a product of two primes. Let's begin with an example.

Example 14.1
Solve the equation $x^{11} \equiv 5 \bmod 47$.

Answer Let x be a solution, so

$$x^{11} \equiv 5 \bmod 47. \tag{14.2}$$

We also know by Fermat's Little Theorem that

$$x^{46} \equiv 1 \bmod 47. \tag{14.3}$$

The idea is to combine (14.2) and (14.3) in a clever way to find x.

The key is to observe that 11 and 46 are coprime, and so, using Proposition 10.3 and Exercise 2 of Chapter 10, we can find positive integers s, t such that $11s - 46t = 1$. So $11s = 1 + 46t$ and $11s \equiv 1 \bmod 46$. Then

$$x \cdot (x^{46})^t = x^{1+46t} = x^{11s} = (x^{11})^s,$$

and so by (14.2) and (14.3),

$$x \equiv 5^s \bmod 47.$$

In fact, using the Euclidean algorithm we find that $1 = 21 \cdot 11 - 5 \cdot 46$, so we take $s = 21, t = 5$. Using the method of successive squares (see Example 13.3), we see that $5^{21} \equiv 15 \bmod 47$.

This shows that $x \pmod{47} = 15$ is the only possible solution. To check that it really is a solution, work backwards in the above calculation: if $x \equiv 5^s \bmod 47$, then

$$x^{11} \equiv 5^{11s} \equiv 5^{1+46t} \equiv 5 \cdot (5^{46})^t \equiv 5 \bmod 47,$$

so this is indeed a solution.

The general case is no harder than this example. Here it is.

PROPOSITION 14.2

Let p be a prime, and let k be a positive integer coprime to $p-1$. Then
(i) there is a positive integer s such that $sk \equiv 1 \bmod (p-1)$, and
(ii) for any $b \in \mathbb{Z}$ not divisible by p, the congruence equation

$$x^k \equiv b \bmod p$$

has a unique solution for x modulo p. This solution is $x \equiv b^s \bmod p$,
where s is as in (i).

PROOF   (i) Since k and $p-1$ are coprime, Proposition 10.3 implies that there are integers s, t such that $sk - t(p-1) = 1$. We wish to take s to be positive; we can do this by adding a multiple of $p-1$ to it, provided we add the same multiple of k to t. (Here we are simply observing that $(s+a(p-1))k - (t+ak)(p-1) = 1$ for any a.) We now have $sk = 1 + t(p-1) \equiv 1 \bmod (p-1)$ with s positive, proving (i).

(ii) Suppose that x is a solution to $x^k \equiv b \bmod p$. Since p does not divide b, it does not divide x, so by Fermat's Little Theorem we have $x^{p-1} \equiv 1 \bmod p$. Hence

$$x \equiv x^{1+t(p-1)} \equiv x^{sk} \equiv (x^k)^s \equiv b^s \bmod p.$$

Hence the only possible solution is $x \equiv b^s \bmod p$, and this is indeed a solution, since

$$(b^s)^k \equiv b^{sk} \equiv b^{1+t(p-1)} \equiv b \cdot (b^{p-1})^t \equiv b \bmod p.$$

∎

A simple modification of the proof enables us to find k^{th} roots modulo a product of two primes. Here is the result, which will be crucial to our discussion of secret codes in the next chapter.

PROPOSITION 14.3

Let p,q be distinct primes, and let k be a positive integer coprime to $(p-1)(q-1)$. Then

(i) there is a positive integer s such that $sk \equiv 1 \mod (p-1)(q-1)$, and

(ii) for any $b \in \mathbb{Z}$ not divisible by p or by q, the congruence equation

$$x^k \equiv b \mod pq$$

has a unique solution for x modulo pq. This solution is $x \equiv b^s \mod pq$, where s is as in (i).

PROOF The proof is just like that of the previous proposition, except that at the beginning we find integers s,t such that

$$sk - t(p-1)(q-1) = 1,$$

and at the end we use Proposition 14.1 instead of Fermat's Little Theorem. ∎

Finding Large Primes

In the next chapter I will introduce you to some very clever secret codes that are used every day for the secure transmission of sensitive information. These codes are based on some of the theory of prime numbers and congruence that we have covered already. For practical use of the codes, one of the basic requirements is the ability to find very large prime numbers using a computer — prime numbers with more than 200 digits are required. So here is a brief discussion of how this is done in practice.

The key is to have a good primality test; in other words, given a very large number N on our computer, we want a method that the computer can quickly apply to tell whether this number N is prime or not. We have already seen earlier in this chapter one idea for such a method, based on the Fermat test: using the method of successive squares, test whether $a^{N-1} \equiv 1 \mod N$ for a reasonable number of values of a. If this is false for any of these values a, then N is not prime by Fermat's Little Theorem. However, if it is true for all the tested values of a, then while it is rather likely that N is prime, it is not definite — there are numbers N (such as 561) for which $a^{N-1} \equiv 1 \mod N$ for all integers a coprime to N, yet N is not prime.

However, there is a neat variant of the Fermat test that does work in practice. It is based on the following simple fact.

PROPOSITION 14.4

Let p be a prime. If a is an integer such that $a^2 \equiv 1 \bmod p$, then $a \equiv \pm 1 \bmod p$.

PROOF Assume that $a^2 \equiv 1 \bmod p$. Then p divides $a^2 - 1$, which is equal to $(a-1)(a+1)$. Hence, by Proposition 10.5(b), p divides either $a-1$ or $a+1$. In other words, either $a \equiv 1 \bmod p$ or $a \equiv -1 \bmod p$. ∎

Here then is the variant of the Fermat test known as *Miller's test*. Let N be an odd positive integer to be tested for primality, and let b be a positive integer less than N (known as the *base*). First, test whether $b^{N-1} \equiv 1 \bmod N$. If this is false we know that N is not prime, and stop; if it is true, we work out $b^{(N-1)/2} \pmod{N}$. If this is not $\pm 1 \pmod{N}$, then N is not prime by Proposition 14.4, and we say that N fails Miller's test with the base b, and stop. If it is $-1 \pmod{N}$, we say that N passes Miller's test with the base b, and stop. And if it is $1 \pmod{N}$, then we repeat: work out $b^{(N-1)/4} \pmod{N}$ (assuming 4 divides $N-1$) — if this is not $\pm 1 \pmod{N}$, then N is not prime and fails Miller's test; if it is $-1 \pmod{N}$, N passes Miller's test; and if it is $1 \pmod{N}$, we repeat, this time working out $b^{(N-1)/8} \pmod{N}$ (assuming 8 divides $N-1$).

We carry on repeating this process. One of the following three things will happen:

(1) at some point we get a value that is not $\pm 1 \pmod{N}$;

(2) at some point we get $-1 \pmod{N}$;

(3) we always get $1 \pmod{N}$ until we run out of powers of 2 to divide $N-1$ by (in other words, we get $b^{(N-1)/2^i} \equiv 1 \bmod N$ for $i = 0, \ldots s$, where $N - 1 = 2^s \cdot m$ with m odd).

If (1) happens, we say that N fails Miller's test with the base b, and if (2) or (3) happens, we say that N passes Miller's test with the base b.

Example 14.2

We've seen (or at least I have told you!) that it is impossible to show that 561 is not prime using the Fermat test (since $a^{560} \equiv 1 \bmod 561$ for all a coprime to 561). But if you do Miller's test with the base 5, you will find at the first step that $5^{(561-1)/2} = 5^{280} \equiv 67 \bmod 561$; hence, 561 fails Miller's test and is not prime.

We know that if N fails Miller's test then it is definitely not prime. But what if N passes the test? Here's the crux: a clever — but not too difficult

— argument shows that if b is chosen at random and N is *not* prime, then the chance that N passes Miller's test with the base b is less than $\frac{1}{4}$. Hence if we do the test with k different bases, the chance N will pass all k tests is less than $(\frac{1}{4})^k$. Taking $k = 100$, say, this chance is $(\frac{1}{4})^{100}$, which is about 10^{-60}, and this is less than the chance that your computer makes an error in its calculations along the way.

To summarise, we now have a powerful, albeit "probabilistic" primality test for a large integer N: pick at random 100 positive integers less than N, and do Miller's test on N with each of these as the base. If N fails any of the tests, it is not prime; and if it passes all of them then N is almost certainly a prime, the chance of getting the answer wrong being less than the chance of a computer error.

For a discussion of this and more powerful tests, see the book by K.H. Rosen listed in the Further Reading at the end of this book.

Exercises for Chapter 14

1. (a) Find $3^{301} \pmod{11}$, $5^{110} \pmod{13}$ and $7^{1388} \pmod{127}$.

   (b) Show that $n^7 - n$ is divisible by 42 for all positive integers n.

2. Let a, b be integers, and let p be a prime not dividing a. Show that the solution of the congruence equation $ax \equiv b \bmod p$ is $x \equiv a^{p-2}b \bmod p$.

   Use this observation to solve the congruence equation $4x \equiv 11 \bmod 19$.

3. Let $N = 561 = 3 \cdot 11 \cdot 17$. Show that $a^{N-1} \equiv 1 \bmod N$ for all integers a coprime to N.

4. Let p be a prime number and k a positive integer.

   (a) Show that if p is odd and x is an integer such that $x^2 \equiv 1 \bmod p^k$, then $x \equiv \pm 1 \bmod p^k$.

   (b) Find the solutions of the congruence equation $x^2 \equiv 1 \bmod 2^k$. (*Hint:* There are different numbers of solutions according to whether $k = 1$, $k = 2$ or $k > 2$.)

5. Show that if p and q are distinct primes, then $p^{q-1} + q^{p-1} \equiv 1 \bmod pq$.

6. The number $(p-1)! \pmod{p}$ came up in our proof of Fermat's Little Theorem, although we didn't need to find it. Calculate $(p-1)! \pmod{p}$ for some small prime numbers p. Find a pattern and make a conjecture.

   Prove your conjecture! (*Hint:* You may find Exercise 6 of Chapter 13 useful.)

7. (a) Solve the congruence equation $x^3 \equiv 2 \bmod 29$.

(b) Find the 7th root of 12 modulo 143 (i.e., solve $x^7 \equiv 12 \bmod 143$).

(c) Find the 11th root of 2 modulo 143.

8. Use Miller's test with a few different bases to try to discover whether 2161 is a prime number. Make sure your answer has a chance of at least 98% of being correct.

9. In a late-night showing of the Spanish cult movie *Teorema Poca de Fermat*, critic Ivor Smallbrain is dreaming that the answer to life, the universe and everything is 1387, provided this number is prime. He tries Fermat's test on 1387, then Miller's test, both with the base 2.

What are the results of Ivor's tests? Has he found the answer to life, the universe and everything?

Chapter 15

Secret Codes

Since time immemorial, people have found it necessary to send information to others in a secure way, so that even if a message is intercepted by a third party, it can be read only by the intended recipient. This is usually achieved by converting the message into a string of numbers or letters according to some rule that only the sender and recipient know. The process of converting the message into such a string is called "encoding," and the process by which the recipient converts the received string back to the original message is called "decoding."

For example, one of the simplest imaginable encoding rules would be to substitute 01 for an A, 02 for a B, 03 for a C, and so on until we get to 26 for a Z. Then the message

SMALLBRAIN IS THE CULPRIT

gets encoded as the string of numbers

$$19130111212020218010914091920080503211216180920. \qquad (15.1)$$

However, anyone with a modicum of intelligence would be able to break this code in a matter of minutes. In view of the millions of Internet transactions, sensitive business and government communications, and so on, that take place every day, some rather more sophisticated ideas are required to ensure the secure transfer of information in today's world. The codes now in use for such purposes — so-called *RSA codes* — are based on some of the theory of prime numbers and congruence developed in the last few chapters. These codes can be explained in just a few pages, which I shall now attempt to do.

Before going into the theory, let me explain the crucial fact that gives these codes their security. Any form of security is based on things that "attackers" are unable to do. In this case, information is being transmitted electronically by computer, so security needs to be based on something that computers can't do quickly.

Let me now explain something that computers can't do quickly. If you take two reasonably large prime numbers, say 1301 and 2089, you can find their product (2717789) using a calculator, in a few seconds. However, if instead you were given the number 2717789 and asked to find its prime factors, it would take you and your calculator much longer. (If you doubt this, try using your calculator to find the prime factors of 127741.) The same kind of observation applies if we bring very powerful computers into play: take two very large primes p and q, say with about 200 digits each — I outlined to you how such primes are found in the last section of Chapter 14 — and calculate pq. Keep p and q secret, but tell the most powerful computer in the world the value of pq. Then, however cleverly that computer is programmed (at least with ideas available today), it will probably take at least several centuries for it to find out what p and q are. In other words, computers find factorization of large numbers "difficult."

RSA Codes: Encoding

Here are the encoding and decoding processes for RSA codes. We choose two large prime numbers p and q, and multiply them together to get $N = pq$. We also find $(p-1)(q-1)$ and choose a large number e which is coprime to $(p-1)(q-1)$. We then make public the numbers N and e, but keep the values of p and q secret. We shall illustrate the discussion with the values

$$p = 37, \ q = 61, \ N = pq = 2257, \ (p-1)(q-1) = 2160, \ e = 11.$$

(In practice, to ensure the security of the code, one uses primes with around 200 digits, but these primes will serve for illustrative purposes.)

The pair (N, e) is called the *public key* of the code. Anyone who wants to send us a message can use the numbers N and e in the following way. First, they convert their message into a string of numbers by some process such as the one given above (01 for A, etc.). They then break up this string into a sequence of numbers with fewer digits than N. For example, with p, q as above, the message (15.1) would be broken up as the sequence

191, 301, 121, 202, 180, 109, 140, 919, 200, 805, 032, 112, 161, 809, 20.

So the message now is a sequence of numbers; call it $m_1, m_2, \ldots, m_k$. The next step is to encode this message. To do this, the sender calculates, for $1 \leq i \leq k$, the value of n_i, where $n_i \equiv m_i^e \bmod N$ and $0 \leq n_i < N$. This can be done quickly on a computer using the method of successive squares described in Example 14.3. The encoded message is the new list of numbers $n_1, n_2, \ldots, n_k$, and this is the message that is sent to us. In the example above, $m_1 = 191$, $N = 2257$ and $e = 11$, and using the method of successive squares, we have

$$191^2 \equiv 369 \bmod 2257, \ \ 191^4 \equiv 741 \bmod 2257, \ \ 191^8 \equiv 630 \bmod 2257,$$

and hence

$$191^{11} \equiv 191^{8+2+1} \equiv 630 \cdot 369 \cdot 191 \equiv 2066 \bmod 2257.$$

Hence $n_1 = 2066$. Similarly, we find that $n_2 = 483$, $n_3 = 914$, $n_4 = 1808$, and so on. So the encoded message is

$$2066, 483, 914, 1808, \ldots$$

Decoding

Once we have received the message, we have to decode it. In other words, given the received list of numbers $n_1, \ldots, n_k$, we need to find the original numbers $m_1, \ldots, m_k$, where $n_i \equiv m_i^e \bmod N$. But m_i is just the solution of the congruence equation $x^e \equiv n_i \bmod N$, and we have seen in Proposition 14.3 how to solve such congruence equations: since e has been chosen to be coprime to $(p-1)(q-1)$, we can use the Euclidean algorithm to find a positive integer d such that $de \equiv 1 \bmod (p-1)(q-1)$. Then the solution to the congruence equation $x^e \equiv n_i \bmod N$ is $x \equiv n_i^d \bmod N$. Of course this solution must be the original number m_i, and hence we recover the original message $m_1, \ldots, m_k$, as desired.

In the above example, $e = 11$ and $(p-1)(q-1) = 2160$. Since $n_1 = 2066$, to recover the first number m_1 of the original message we need to solve the congruence equation $x^{11} \equiv 2066 \bmod 2257$. Use of the Euclidean algorithm shows that $1 = 1571 \cdot 11 - 8 \cdot 2160$, so we take $d = 1571$. We then use successive squares to calculate that $2066^{1571} \equiv 191 \bmod 2257$ and hence recover the first number, $m_1 = 191$, of the original message. Similarly, we recover $m_2, \ldots, m_k$.

To summarise: to encode a message, we raise each of its listed numbers to the power e (e for "encode") and work out the answer module N; and to decode a received message, we raise each of its listed numbers to the power d (d for "decode") and work this out modulo N.

Now let's decode a new message (still keeping the above values of p, q and e). You are relaxing on your hotel balcony in Monte Carlo, sipping a pina colada, when your laptop beeps and the following strange message arrives:

$$763, 28, 1034, 559, 2067, 2028, 798.$$

At once you spring into action, getting your laptop to solve the congruence equations

$$x^{11} \equiv 763 \bmod 2257 \;\rightarrow\; x \equiv 763^{1571} \bmod 2257 \;\rightarrow\; x\,(\bmod\,2257) = 251$$
$$x^{11} \equiv 28 \bmod 2257 \;\;\;\;\rightarrow\; x \equiv 28^{1571} \bmod 2257 \;\;\;\;\rightarrow\; x\,(\bmod\,2257) = 521$$
$$x^{11} \equiv 1034 \bmod 2257 \;\rightarrow\; x \equiv 1034^{1571} \bmod 2257 \;\rightarrow\; x\,(\bmod\,2257) = 180$$
$$x^{11} \equiv 559 \bmod 2257 \;\;\rightarrow\; x \equiv 559^{1571} \bmod 2257 \;\;\rightarrow\; x\,(\bmod\,2257) = 506$$
$$x^{11} \equiv 2067 \bmod 2257 \;\rightarrow\; x \equiv 2076^{1571} \bmod 2257 \;\rightarrow\; x\,(\bmod\,2257) = 091$$
$$x^{11} \equiv 2028 \bmod 2257 \;\rightarrow\; x \equiv 2028^{1571} \bmod 2257 \;\rightarrow\; x\,(\bmod\,2257) = 805$$
$$x^{11} \equiv 798 \bmod 2257 \;\;\rightarrow\; x \equiv 798^{1571} \bmod 2257 \;\;\rightarrow\; x\,(\bmod\,2257) = 04.$$

This gives the decoded message as the following string of numbers:

$$251, 521, 180, 506, 091, 805, 04.$$

Using the original substitutions, 01 for A, 02 for B, etc., you finally translate this into the urgent message:

YOUREFIRED

Unruffled, you saunter back to your drink. You'd been planning to take up that managing directorship offer anyway . . .

Security

How secure is the RSA code described above? In other words, if an encoded message is intercepted by a third party (who knows the publicly available values of N and e), how easy is it for them to decode the message? Well, at present, the only known way to decode is first to find the value of $(p-1)(q-1)$ and then to calculate d and use the decoding method above, working out d^{th} powers modulo N.

However, $(p-1)(q-1) = pq - p - q + 1 = N - (p+q) + 1$, so if we know N and can find $(p-1)(q-1)$, then we can also find $p+q$. But then we can also find p and q, since they are the roots of the quadratic equation $x^2 - (p+q)x + N = 0$.

In other words, in order to be able to decode messages, a third party needs to be able to find the prime factors p and q of N. But as I explained before starting the description of RSA codes, if p and q are very large primes — both with about 200 digits — then no computer on earth will be able to find p and q, given only their product. In other words, the code is secure!

Even if increases in computer power enable us to factorize products of two 200-digit primes in the future, it will be a simple matter to retain security just by increasing the number of digits in our primes p and q. What is not clear is whether someone in the future will discover a clever new method of computer factorization of large numbers that makes all these codes insecure.

A Little History

The remarkable idea that it might be possible to design a code where knowledge of the encoding process does not mean that one can decode messages was proposed by Diffie and Hellman in 1976, and a year later Rivest, Shamir and Adleman found such codes, namely the ones described above, which are known as RSA codes in honour of their discoverers. A few years ago, however, some new facts came to light from the Government Communications Headquarters (GCHQ) in Cheltenham in the UK, and it turned out that these codes had been discovered there several years earlier by Ellis, Cocks and Williamson. These discoveries could not be publicised at the time because of the classified secret nature of work at GCHQ. For an interesting account of this and many other aspects of secret codes through the ages, read the book by Simon Singh listed in the Further Reading at the end of this book.

Exercises for Chapter 15

1. Find the primes p and q, given that $pq = 18779$ and $(p-1)(q-1) = 18480$.

2. (a) Encode the message WHERE ARE YOU using the public key $(N, e) = (143, 11)$.

   (b) You intercept the encoded answer. Here it is:

   $$12, 59, 14, 114, 59, 14.$$

   Brilliantly crack this code and decipher this message.

3. Critic Ivor Smallbrain has been given the honour of being given a chance to nominate the best film of all time by the Oscar committee. He has to send his nomination to them using the RSA code with public key $(N, e) = (1081, 25)$. The nomination will only be accepted if it remains secret until the ceremony.

   Unfortunately, arch-rival Greta Picture intercepts Ivor's message. Here it is:

   $$23, 930, 228, 632.$$

   Greta offers a large sum to anyone who can crack the code and decipher Ivor's message.

   Can you improve your bank balance and disappoint Ivor?

Chapter 16

Counting and Choosing

Mathematics has many tools for counting. We shall give some simple methods in this chapter. These lead us to binomial coefficients and then to the Binomial Theorem and the Multinomial Theorem.

Let us begin with an example.

Example 16.1
A security system uses passwords consisting of two letters followed by two digits. How many different passwords are possible?

Answer The number of choices for each letter is 26 and for each digit is 10. We claim the answer is the product of these numbers, namely

$$26 \times 26 \times 10 \times 10.$$

Here is the justification for this claim. Let N be the number of possible passwords, and let a typical password be $\alpha\beta\gamma\delta$, where α, β are letters and γ, δ are digits. For each choice of $\alpha\beta\gamma$, there are 10 passwords $\alpha\beta\gamma\delta$ (one for each of the 10 possibilities for δ). Thus,

$$N = 10 \times \text{number of choices for } \alpha\beta\gamma.$$

Likewise, for each choice of $\alpha\beta$, there are 10 possibilities for $\alpha\beta\gamma$, so

$$N = 10 \times (10 \times \text{number of choices for } \alpha\beta).$$

For each choice of α there are 26 possibilities for $\alpha\beta$, so

$$N = 10 \times 10 \times (26 \times \text{number of choices for } \alpha) = 10 \times 10 \times 26 \times 26.$$

The argument used in the above example shows the following.

THEOREM 16.1 (Multiplication Principle)
Let P be a process which consists of n stages, and suppose that for each r, the rth stage can be carried out in a_r ways. Then P can be carried out in $a_1 a_2 \ldots a_n$ ways.

In the above example, there are four stages, with $a_1 = a_2 = 26, a_3 = a_4 = 10$. Here is another example that is not quite so simple.

Example 16.2
Using the digits $1, 2, \ldots, 9$, how many even two-digit numbers are there with two different digits?

Answer Let us consider the following two-stage process to pick such an even integer. The first stage is to choose the *second* digit (2, 4, 6 or 8); it can be done in four ways. The second stage is to choose the first digit, which can be done in eight ways. Hence, by the Multiplication Principle, the answer to the question is 32.

Notice that we would have had trouble if we had carried out the process the other way around and first chosen the *first* digit, because then the number of ways of choosing the second digit would depend on whether the first digit was even or odd.

Here is another application of the Multiplication Principle.

PROPOSITION 16.1
Let S be a set consisting of n elements. Then the number of different arrangements of the elements of S in order is n!
(Recall that $n! = n(n-1)(n-2) \cdots 2 \cdot 1$.)

For example, if $S = \{a, b, c\}$ then the different arrangements of the elements in order are
$$abc, \ acb, \ bac, \ bca, \ cab, \ cba.$$
As predicted by the proposition, there are $6 = 3!$ of them.

PROOF   Choosing an arrangement is an *n*-stage process. First choose the first element (*n* possibilities); then choose the second element — this can be any of the remaining $n - 1$ elements, so there are $n - 1$ possibilities; then the third, for which there are $n - 2$ possibilities; and so on. Hence, by the Multiplication Principle, the total number of arrangements is $n(n-1)(n-2) \ldots 2 \cdot 1 = n!$. ∎

Binomial Coefficients

We now introduce some numbers that are especially useful in counting arguments.

DEFINITION   *Let n be a positive integer and r an integer such that $0 \leq r \leq n$. Define*

$$\binom{n}{r}$$

(called "n choose r") to be the number of r-element subsets of $\{1,2,\dots,n\}$.

For example, the 2-element subsets of $\{1,2,3,4\}$ are $\{1,2\}$, $\{1,3\}$, $\{1,4\}$, $\{2,3\}$, $\{2,4\}$, $\{3,4\}$, and so

$$\binom{4}{2} = 6.$$

PROPOSITION 16.2
We have

$$\binom{n}{r} = \frac{n!}{r!(n-r)!}.$$

PROOF   Let $S = \{1,2,\dots,n\}$ and count the arrangements of S in order as follows:

Stage 1: Choose an r-element subset T of S: there are $\binom{n}{r}$ choices.
Stage 2: Choose an arrangement of T: by Proposition 16.1 there are $r!$ choices.
Stage 3: Choose an arrangement of the remaining $n-r$ elements of S: there are $(n-r)!$ choices.

By the Multiplication Principle, the total number of arrangements of S is equal to the product of these three numbers. Hence, by Proposition 16.1,

$$n! = \binom{n}{r} \times r! \times (n-r)!$$

and the result follows from this.   ∎

Another way of expressing the conclusion of Proposition 16.2 is

$$\binom{n}{r} = \frac{n(n-1)\dots(n-r+1)}{r(r-1)\dots 2 \cdot 1}.$$

Some useful particular cases of this are:

$$\binom{n}{0} = \binom{n}{n} = 1$$

(there is only one 0-element and one n-element subset of S!), and

$$\binom{n}{1} = n, \quad \binom{n}{2} = \frac{n(n-1)}{2}, \quad \binom{n}{3} = \frac{n(n-1)(n-2)}{6}.$$

Example 16.3
Liebeck has taught the same course for the last 16 years, and tells three jokes each year. He never tells the same set of three jokes twice. At least how many jokes does Liebeck know? When will he have to tell a new one?

Answer Suppose Liebeck knows n jokes. Then $\binom{n}{3}$ must be at least 16. Since $\binom{5}{3} = 10$ and $\binom{6}{3} = 20$, it follows that $n \geq 6$. So Liebeck knows at least 6 jokes and will have to tell a seventh in 5 years' time (i.e., in the 21st year of giving the course — assuming he has not dropped dead by then).

Example 16.4
How many solutions are there of the equation $x+y+z = 10$, where x, y, z are non-negative integers ?

Answer Here's a clever way to approach this. Think of a solution to $x+y+z = 10$ (x, y, z non-negative integers) as a string of ten 1's and two 0's: start the string with x 1's, then a 0, then y 1's, then a 0, then z 1's. (For example, the solution $x = 4, y = 1, z = 5$ is represented by the string 111101011111.) To specify a solution, we write down the 12 symbols, and the only choice is where we put the two 0's. So the total number of solutions is $\binom{12}{2} = 66$.

The numbers $\binom{n}{r}$ are known as *binomial coefficients*. This is because of the following famous theorem.

THEOREM 16.2 Binomial Theorem
Let n be a positive integer, and let a, b be real numbers. Then

$$(a+b)^n = \sum_{r=0}^{n} \binom{n}{r} a^{n-r} b^r$$

$$= a^n + na^{n-1}b + \binom{n}{2} a^{n-2}b^2 + \cdots$$

$$+ \binom{n}{r} a^{n-r} b^r + \cdots + \binom{n}{n-1} ab^{n-1} + b^n.$$

PROOF   Consider

$$(a+b)^n = (a+b)(a+b)\ldots(a+b).$$

When we multiply this out to get a term $a^{n-r}b^r$, we choose the b from r of the brackets and the a from the other $n-r$ brackets. So the number of ways of getting $a^{n-r}b^r$ is the number of ways of choosing r brackets from n, and hence is $\binom{n}{r}$. In other words, the coefficient of $a^{n-r}b^r$ is $\binom{n}{r}$. The theorem follows.   ∎

In Exercise 9 at the end of the chapter, you are asked to provide an alternative proof of the Binomial Thoerem by induction.

Here are the first few cases of the Binomial Theorem:

$$(a+b)^0 = 1$$
$$(a+b)^1 = a+b$$
$$(a+b)^2 = a^2 + 2ab + b^2$$
$$(a+b)^3 = a^3 + 3a^2b + 3ab^2 + b^3$$
$$(a+b)^4 = a^4 + 4a^3b + 6a^2b^2 + 4ab^3 + b^4$$
$$(a+b)^5 = a^5 + 5a^4b + 10a^3b^2 + 10a^2b^3 + 5ab^4 + b^5.$$

There are one or two patterns to observe about this. First, each expression is symmetrical about the centre; this is due to the fact, obvious from Proposition 16.2, that

$$\binom{n}{r} = \binom{n}{n-r}$$

(for example, $\binom{5}{2} = \binom{5}{3} = 10$). Rather less obvious is the fact that if we write down the coefficients in the above expressions in the following array, known as *Pascal's triangle:*

$$
\begin{array}{cccccc}
1 & & & & & \\
1 & 1 & & & & \\
1 & 2 & 1 & & & \\
1 & 3 & 3 & 1 & & \\
1 & 4 & 6 & 4 & 1 & \\
1 & 5 & 10 & 10 & 5 & 1
\end{array}
$$

then you can see that each entry is the sum of the one above it and the one to the left of that. This is explained by the equality

$$\binom{n+1}{r} = \binom{n}{r} + \binom{n}{r-1},$$

which you are asked to prove in Exercise 8 at the end of the chapter.

Putting $a = 1$ and $b = x$ in the Binomial Theorem, we obtain the following consequence.

PROPOSITION 16.3
For any positive integer n,

$$(1+x)^n = \sum_{r=0}^{n} \binom{n}{r} x^r.$$

Putting $x = \pm 1$ in this, we get the interesting equalities

$$\sum_{r=0}^{n} \binom{n}{r} = 2^n, \qquad \sum_{r=0}^{n} (-1)^r \binom{n}{r} = 0.$$

The second of these equalities gives the following, which will be useful in the next chapter:

$$\sum_{r=1}^{n} (-1)^{r-1} \binom{n}{r} = \binom{n}{0} = 1. \tag{16.1}$$

Ordered Selections

Suppose we have a set S of n elements. We know that the number of ways of choosing a subset of S of size r is equal to $\binom{n}{r}$. But there might be different ways we want to choose our elements — for example, we may care about the order in which we select them, or we may want to allow repetitions in our selection. Here's an example to illustrate this.

Example 16.5
Let $S = \{a,b,c,d,e\}$. A "word" is an ordered selection of letters from S — for example *abc*, *cba* or *bbcbe*.

(1) How many three-letter words are there?

(2) How many three-letter words are there with distinct letters?

Answer (1) There are 5 choices for the first letter, 5 for the second and 5 for the third. So the total number of words is $5^3 = 125$.

(2) As the letters are distinct there are 5 choices for the first letter, 4 for the second and 3 for the third. So the total number of words with distinct letters is $5 \cdot 4 \cdot 3 = 60$.

PROPOSITION 16.4
Let S be a set of n elements.

(1) *The number of ordered selections of r elements of S, allowing repetitions, is equal to n^r.*

(2) *The number of ordered selections of r distinct elements of S is equal to $n(n-1)\cdots(n-r+1)$.*

PROOF (1) An ordered selection of r elements of S is a process of r stages: Stage 1 is to choose the first element, Stage 2 the second, and so on. There are n choices at each stage. Hence the total number of selections is n^r.

(2) Similarly, an ordered selection of r distinct elements of S is a process of r stages in which there are n choices at the first stage, $n-1$ at the second, and so on until the rth stage at which there are $n-r+1$ choices. Hence the total number of selections is $n(n-1)\cdots(n-r+1)$.

We will write $P(n,r)$ for the number of ordered selections of r distinct elements from a set of size n, so that

$$P(n,r) = n(n-1)\cdots(n-r+1).$$

Note that $P(n,r) = \frac{n!}{(n-r)!} = r!\binom{n}{r}$.

Example 16.6
The Birthday Paradox Suppose there are r people in a room. Let's work out the chance that they all have birthdays on different dates. (For simplicity we will assume that all years have 365 days and that each date is equally likely as a birthday.) By Proposition 16.4, the number of ordered selections of r *different* dates is equal to $P(365,r)$, while the total number of ordered selections of r dates is 365^r. Hence the chance that they all have different birthdays is

$$\frac{P(365,r)}{365^r} = \frac{365\cdot364\cdots(365-r+1)}{365\cdot365\cdots365}.$$

This ratio can be calculated for various values of r. For example, for $r = 23$ it is about 0.493, and for $r = 70$ it is less than 0.001. This means that in a room with 23 people, the chance that two of them have the *same* birthday is more than 50%; and with 70 people, the chance that two have the same birthday is more than 99.9%! These facts may be surprising at first sight, and the one about the 23 people is called the Birthday Paradox.

Multinomial Coefficients

Suppose eight students (call them $1,2,3,\ldots,8$) are to be assigned to three projects (call them A, B, C); project A requires 4 students, project B requires 2, and project C also requires 2. In how many ways can the students be assigned to the projects?

To answer this, we could list all possible assignments like this:

A	B	C
1,2,3,4	5,6	7,8
1,2,3,4	7,8	5,6
1,2,3,4	5,7	6,8
1,2,3,4	6,8	5,7
1,2,3,5	4,6	7,8
.	.	.
.	.	.

However, the deadline for the projects will probably have passed by the time we have finished writing down the complete list (in fact there are 420 possible assignments). We need a nice way of counting such things.

Each assignment is what is called an *ordered partition* of the set $\{1,2,\ldots,8\}$ into subsets A,B,C of sizes 4, 2, 2. Here is the general definition of such a thing.

DEFINITION   *Let n be a positive integer, and let $S = \{1,2,\ldots,n\}$. A partition of S is a collection of subsets $S_1,\ldots,S_k$ such that each element of S lies in exactly one of these subsets. The partition is* ordered *if we take account of the order in which the subsets are written.*

The point about the order is that, for instance in the above example, the ordered partition

$$\{1,2,3,4\} \quad \{5,6\} \quad \{7,8\}$$

is different from the ordered partition

$$\{1,2,3,4\} \quad \{7,8\} \quad \{5,6\}$$

even though the subsets involved are the same in both cases.

If $r_1,\ldots,r_k$ are non-negative integers such that $n = r_1 + \cdots + r_k$, we denote the total number of ordered partitions of $S = \{1,2,\ldots,n\}$ into subsets $S_1,\ldots,S_k$ of sizes $r_1,\ldots,r_k$ by the symbol

$$\binom{n}{r_1,\ldots,r_k}.$$

Example 16.7

(1) The number of possible project assignments in the example above is

$$\binom{8}{4,2,2}.$$

(2) If Alfred, Barney, Cedric and Dugald play bridge, the total number of different possible hands that can be dealt is

$$\binom{52}{13,13,13,13}.$$

(3) It is rather clear that

$$\binom{n}{r,n-r} = \binom{n}{r}.$$

PROPOSITION 16.5

We have

$$\binom{n}{r_1,\ldots,r_k} = \frac{n!}{r_1!r_2!\ldots r_k!}.$$

PROOF   We count the $n!$ arrangements of $S = \{1,2,\ldots,n\}$ in stages as follows:

Stage 0: Choose an ordered partition of S into subsets $S_1,\ldots,S_k$ of sizes $r_1,\ldots,r_k$; the number of ways of doing this is

$$\binom{n}{r_1,\ldots,r_k}.$$

Stage 1: Choose an arrangement of S_1: there are $r_1!$ choices.

Stage 2: Choose an arrangement of S_2: there are $r_2!$ choices.

And so on, until

Stage k: Choose an arrangement of S_k: there are $r_k!$ choices.

By the Multiplication Principle, we conclude that

$$n! = \binom{n}{r_1,\ldots,r_k}r_1!\ldots r_k!$$

The result follows.   ∎

Example 16.8
(1) The number of project assignments in the first example is

$$\binom{8}{4,2,2} = \frac{8!}{4!2!2!} = 420.$$

(2) The total number of bridge hands, namely

$$\binom{52}{13,13,13,13},$$

is approximately 5.365×10^{28}; quite a large number.

(3) What is the total number of ways of arranging the letters of the word MISSISSIPPI? Well, each arrangement corresponds to an ordered partition of the 11 positions for the letters, with subsets of sizes 4 (for the I's), 4 (for the S's), 2 (for the P's) and 1 (for the M). So the total number of arrangements is

$$\binom{11}{4,4,2,1} = 34650.$$

The numbers $\binom{n}{r_1,\ldots,r_k}$ are called *multinomial coefficients,* for the following reason.

THEOREM 16.3 Multinomial Theorem
Let n be a positive integer, and let $x_1,\ldots,x_k$ be real numbers. Then the expansion of $(x_1 + \cdots + x_k)^n$ is the sum of all terms of the form

$$\binom{n}{r_1,\ldots,r_k} x_1^{r_1} \ldots x_k^{r_k}$$

where $r_1,\ldots,r_k$ are non-negative integers such that $r_1 + \cdots + r_k = n$.

PROOF   Consider

$$(x_1 + \cdots + x_k)^n = (x_1 + \cdots + x_k)(x_1 + \cdots + x_k)\ldots(x_1 + \cdots + x_k).$$

In expanding this, we get a term $x_1^{r_1} \ldots x_k^{r_k}$ by choosing x_1 from r_1 of the brackets, x_2 from r_2 brackets, and so on. The number of ways of doing this is

$$\binom{n}{r_1,\ldots,r_k},$$

so this is the coefficient of $x_1^{r_1} \ldots x_k^{r_k}$ in the expansion. ∎

Example 16.9

(1) The expansion of $(x+y+z)^3$ is

$$(x+y+z)^3 = x^3 + y^3 + z^3 + 3x^2y + 3xy^2 + 3x^2z + 3xz^2 + 3y^2z + 3yz^2 + 6xyz.$$

(2) The coefficient of $x^2y^3z^2$ in the expansion of $(x+y+z)^7$ is

$$\binom{7}{2,3,2} = \frac{7!}{2!3!2!} = 210.$$

(3) Find the coefficient of x^3 in the expansion of $(1 - \frac{1}{x^3} + 2x^2)^5$.

Answer A typical term in this expansion is

$$\binom{5}{a,b,c} \cdot 1^a \cdot (\frac{-1}{x^3})^b \cdot (2x^2)^c$$

where $a+b+c = 5$ (and $a,b,c \geq 0$). To make this a term in x^3, we need

$$-3b + 2c = 3 \quad \text{and} \quad a+b+c = 5.$$

From the first equation, 3 divides c, so $c = 0$ or 3. If $c = 0$ then $b = -1$, which is impossible. Hence $c = 3$, and it follows that $a = 1, b = 1$. Thus there is just one term in x^3, namely

$$\binom{5}{1,1,3}(\frac{-1}{x^3})(2x^2)^3 = -160x^3.$$

In other words, the coefficient is -160.

Exercises for Chapter 16

1. Evaluate the binomial coefficients $\binom{8}{3}$ and $\binom{15}{5}$.

2. Liebeck, Einstein and Hawking pinch their jokes from a joke book which contains 12 jokes. Each year Liebeck tells six jokes, Einstein tells four and Hawking tells two (and everyone tells different jokes). For how many years can they go on, never telling the same three sets of jokes?

3. (a) How many solutions are there of the equation $x+y+z+t = 14$, where x,y,z,t are non-negative integers? (Hint: see Example 16.4.)

   (b) How many solutions are there of the equation $x+y+z+t = 14$, where x,y,z,t are positive integers and $t \leq 8$?

   (c) Let $c_1,\ldots,c_r$ be integers, and let N be an integer such that $N \geq \sum_1^r c_i$. Find, in terms of N, r and $c_1,\ldots,c_r$, a formula for the number of solutions of the equation $x_1 + \cdots + x_r = N$, where the x_i are integers and $x_i \geq c_i$ for all i.

4. Josephine lives in the lovely city of Blockville. Every day Josephine walks from her home to Blockville High School, which is located 10 blocks east and 14 blocks north from home. She always takes a shortest walk of 24 blocks.

(a) How many different walks are possible?

(b) 4 blocks east and 5 blocks north of Josephine's home lives Jemima, her best friend. How many different walks to school are possible for Josephine if she meets Jemima at Jemima's home on the way?

(c) There is a park 3 blocks east and 6 blocks north of Jemima's home. How many walks to school are possible for Josephine if she meets Jemima at Jemima's home and they then stop in the park on the way?

5. (a) How many words of ten or fewer letters can be formed using the alphabet $\{a,b\}$?

(b) Using the alphabet $\{a,b,c,d,e,f\}$, how many six letter words are there that use all six letters, in which no two of the letters a,b,c occur consecutively?

6. (a) Find the number of arrangements of the set $\{1,2,\ldots,n\}$ in which the numbers 1,2 appear as neighbours.

(b) Let $n \geq 5$. Find the number of arrangements of the set $\{1,2,\ldots,n\}$ in which the numbers 1,2,3 appear as neighbours in order, and so do the numbers 4,5.

7. Liebeck has n steaks and is surrounded by n hungry wolves. He throws each of the steaks to a random wolf. What is the chance that

(i) every wolf gets a steak?

(ii) exactly one wolf does not get a steak?

(iii) Liebeck gets eaten, in the case where $n = 7$?

8. (a) Prove that
$$\binom{n+1}{r} = \binom{n}{r} + \binom{n}{r-1}.$$

(b) Prove that for any positive integer n,
$$3^n = \sum_{k=0}^{n} \binom{n}{k} 2^k.$$

9. Give a proof of the Binomial Theorem 16.2 by induction on n.

10. n points are placed on a circle, and each pair of points is joined by a straight line. The points are chosen so that no three of these lines pass through the same point. Let r_n be the number of regions into which the interior of the circle is divided.

Draw pictures to calculate r_n for some small values of n.

Conjecture a formula for r_n in terms of n.

11. Three tickets are chosen from a set of 100 tickets numbered $1, 2, 3, \ldots,$ 100. Find the number of choices such that the numbers on the three tickets are

(a) in arithmetic progression (i.e., $a, a+d, a+2d$ for some a, d)

(b) in geometric progression (i.e., a, ar, ar^2 for some a, r).

12. The digits $1, 2, 3, 4, 5, 6$ are written down in some order to form a six-digit number.

(a) How many such six-digit numbers are there altogether?

(b) How many such numbers are even?

(c) How many are divisible by 4?

(d) How many are divisible by 8? (*Hint:* First show that the remainder on dividing a six-digit number $abcdef$ by 8 is $4d + 2e + f$.)

13. (a) Find the coefficient of x^{15} in $(1+x)^{18}$.

(b) Find the coefficient of x^4 in $(2x^3 - \frac{1}{x^2})^8$.

(c) Find the constant term in the expansion of $(y + x^2 - \frac{1}{xy})^{10}$.

14. The rules of a lottery are as follows: You select 10 numbers between 1 and 50. On lottery night, celebrity mathematician Richard Thomas chooses at random 6 "correct" numbers. If your 10 numbers include all 6 correct ones, you win.

Work out your chance of winning the lottery.

15. Here's another way to prove Fermat's Little Theorem. Let p be a prime number.

(a) Show that if r, s are positive integers such that s divides r, p divides r and p does not divide s, then p divides $\frac{r}{s}$.

(b) Deduce that p divides the binomial coefficient $\binom{p}{k}$ for any k such that $1 \le k \le p-1$.

(c) Now use the Binomial Theorem to prove by induction on n that p divides $n^p - n$ for all positive integers n. Hence, deduce Fermat's Little Theorem.

16. At a party with six rather decisive people, any two people either like each other or dislike each other. Prove that at this party, either

   (i) there are three people all of whom like each other, or

   (ii) there are three people, all of whom dislike each other.

   Show that it is possible to have a party with five (decisive) people where neither (i) nor (ii) holds.

17. Prove that if r_n is as in Question 10, then for any n,

$$r_n = 1 + \binom{n}{2} + \binom{n}{4}.$$

   Was your conjecture in Question 10 correct?

18. The other day, critic Ivor Smallbrain gave a lecture to an audience consisting of five mathematicians. Each mathematician fell asleep exactly twice during the lecture. For each pair of mathematicians, there was a moment during the lecture when they were both asleep. Prove that there was a moment when three of the mathematicians were simultaneously asleep.

Chapter 17

More on Sets

In this chapter we develop a little of the theory of sets. After some notation and a few elementary results, we present the "Inclusion–Exclusion Principle," which is another useful counting method to add to those of the previous chapter.

Unions and Intersections

We begin with a couple of definitions.

DEFINITION   *Let A and B be sets. The* union *of A and B, written $A \cup B$, is the set consisting of all elements that lie in either A or B (or both). Symbolically,*

$$A \cup B = \{x \mid x \in A \ or \ x \in B\}.$$

The intersection *of A and B, written $A \cap B$, is the set consisting of all elements that lie in both A and B; thus*

$$A \cap B = \{x \mid x \in A \ and \ x \in B\}.$$

Example 17.1
   (1) If $A = \{1,2,3\}$ and $B = \{2,4\}$, then $A \cup B = \{1,2,3,4\}$ and $A \cap B = \{2\}$.
   (2) Let $A = \{n \mid n \in \mathbb{Z}, n \geq 0\}$ and $B = \{n \mid n \in \mathbb{Z}, n \leq 0\}$. Then $A \cup B = \mathbb{Z}$ and $A \cap B = \{0\}$.

We say that A and B are *disjoint* sets if they have no elements in common — that is, if $A \cap B = \emptyset$, the empty set.

Recall from Chapter 1 that the notation $A \subseteq B$ means that A is a subset of B, i.e., every element of A lies in B, which is to say $x \in A \Rightarrow x \in B$. Also, we define $A = B$ to mean that A and B have exactly the same elements. Other ways of expressing $A = B$ are: both $A \subseteq B$ and $B \subseteq A$; or $x \in A \Leftrightarrow x \in B$. As a further piece of notation, we write $A \subset B$ to mean that A is a subset of B and $A \neq B$; when $A \subset B$ we say A is a *proper* subset of B.

The "algebra of sets" consists of general results involving sets, unions and intersections. Such results are usually pretty uninteresting. Here is one.

PROPOSITION 17.1
Let A, B, C be sets. Then
$$A \cap (B \cup C) = (A \cap B) \cup (A \cap C).$$

PROOF   This just involves keeping careful track of the definitions:

$$x \in A \cap (B \cup C) \Leftrightarrow x \in A \text{ and } x \in (B \text{ or } C)$$
$$\Leftrightarrow (x \in A \text{ and } x \in B) \text{ or } (x \in A \text{ and } x \in C)$$
$$\Leftrightarrow x \in (A \cap B) \cup (A \cap C).$$

Hence, $A \cap (B \cup C) = (A \cap B) \cup (A \cap C).$ ∎

More examples of results in the algebra of sets can be found in the exercises at the end of the chapter.

We can extend the definitions of union and intersection to many sets: if $A_1, A_2, \ldots, A_n$ are sets, their union and intersection are defined as

$$A_1 \cup A_2 \cup \ldots \cup A_n = \{x \mid x \in A_i \text{ for some } i\},$$
$$A_1 \cap A_2 \cap \ldots \cap A_n = \{x \mid x \in A_i \text{ for all } i\}.$$

We sometimes use the more concise notation

$$A_1 \cup \ldots \cup A_n = \bigcup_{i=1}^{n} A_i, \quad A_1 \cap \ldots \cap A_n = \bigcap_{i=1}^{n} A_i.$$

Likewise, if we have an infinite collection of sets $A_1, A_2, A_3, \ldots$, their union and intersection are defined as

$$\bigcup_{i=1}^{\infty} A_i = \{x \mid x \in A_i \text{ for some } i\}, \quad \bigcap_{i=1}^{\infty} A_i = \{x \mid x \in A_i \text{ for all } i\}.$$

Example 17.2
For $i \geq 1$ let $A_i = \{x \mid x \in \mathbb{Z}, x \geq i\}$. Then

$$\bigcup_{i=1}^{\infty} A_i = \mathbb{N}, \quad \bigcap_{i=1}^{\infty} A_i = \emptyset.$$

If A, B are sets, their *difference* is defined to be the set

$$A - B = \{x \,|\, x \in A \text{ and } x \notin B\}.$$

For example, if $A = \{x \,|\, x \in \mathbb{R}, 0 \leq x \leq 1\}$ and $B = \mathbb{Q}$, then $A - B$ is the set of irrationals between 0 and 1.

Cartesian Products

Cartesian products give a way of constructing new sets from old.

DEFINITION   *Let A, B be sets. The Cartesian product of A and B, written $A \times B$, is the set consisting of all symbols of the form (a, b) with $a \in A, b \in B$. Such a symbol (a, b) is called an ordered pair of elements of A and B. Two ordered pairs $(a, b), (a', b')$ are deemed to be equal if and only if both $a = a'$ and $b = b'$.*

For example, if $A = \{1, 2\}$ and $B = \{1, 4, 5\}$, then $A \times B$ consists of the six ordered pairs

$$(1, 1), (1, 4), (1, 5), (2, 1), (2, 4), (2, 5).$$

As another example, when $A = B = \mathbb{R}$, the Cartesian product is $\mathbb{R} \times \mathbb{R}$, which consists of all ordered pairs (x, y) $(x, y \in \mathbb{R})$, commonly known as coordinate pairs of points in the plane.

We can also form the Cartesian product of more than two sets in a similar way: if $A_1, A_2, \ldots, A_n$ are sets, their Cartesian product is defined to be the set $A_1 \times A_2 \times \ldots \times A_n$ consisting of all symbols of the form $(a_1, a_2, \ldots, a_n)$, where $a_i \in A_i$ for all i. Such symbols are called *n-tuples* of elements of $A_1, \ldots, A_n$.

The Inclusion–Exclusion Principle

Logically enough, we call a set S a *finite* set if it has only a finite number of elements. If S has n elements, we write $|S| = n$. If a set is not finite, it is said to be an *infinite* set.

For example, if $S = \{1, -3, \sqrt{2}\}$, then S is finite and $|S| = 3$. And $\mathbb{Z}$ is an infinite set.

Here is a useful result about finite sets.

PROPOSITION 17.2
If A and B are finite sets, then

$$|A \cup B| = |A| + |B| - |A \cap B|.$$

PROOF   Let $|A \cap B| = k$, say $A \cap B = \{x_1, \ldots, x_k\}$. These elements, and no others, belong to both A and B, so we can write

$$A = \{x_1, \ldots, x_k, a_1, \ldots, a_l\}, \qquad B = \{x_1, \ldots, x_k, b_1, \ldots, b_m\},$$

where $|A| = k + l$, $|B| = k + m$. Then

$$A \cup B = \{x_1, \ldots, x_k, a_1, \ldots, a_l, b_1, \ldots, b_m\},$$

and all these elements are different, so

$$\begin{aligned} |A \cup B| = k + l + m &= (k + l) + (k + m) - k \\ &= |A| + |B| - |A \cap B|. \quad \blacksquare \end{aligned}$$

Example 17.3
Out of a total of 30 students, 19 are doing mathematics, 17 are doing music and 10 are doing both. How many are doing neither?

Answer Let A be the set doing mathematics and B the set doing music. Then

$$|A| = 19, |B| = 17, |A \cap B| = 10.$$

Hence, Proposition 17.2 gives $|A \cup B| = 19 + 17 - 10 = 26$. Since there are 30 students in all, there are therefore 4 doing neither mathematics nor music.

Proposition 17.2 can be generalized to give a formula for the size of the union of any number of finite sets. First let's consider the case of three sets A, B, C. Here the formula is

$$|A \cup B \cup C| = |A| + |B| + |C| - |A \cap B| - |A \cap C| - |B \cap C| + |A \cap B \cap C|. \quad (17.1)$$

This can be proved in similar fashion to Proposition 17.2, but let me present a slightly more concise argument. Consider an element x of $A \cup B \cup C$. I will argue that x contributes precisely 1 to the right-hand side of the equation (17.1). If x belongs to exactly one of the sets A, B, C, say to A, then it contributes only to the term $|A|$ in (17.1). If x lies in two of the sets, say in A and B, then it contributes 1 to $|A|$, 1 to $|B|$ and -1 to $-|A \cap B|$, hence a total of 1. And if x lies in all three sets, it contributes 1 to $|A|$, $|B|$ and $|C|$; -1 to $-|A \cap B|$, $-|A \cap C|$ and $-|B \cap C|$; and 1 to $|A \cap B \cap C|$, making a total of 1 overall. Hence as claimed, each element $x \in A \cup B \cup C$ contributes precisely 1 to the right-hand side of the equation (17.1), so (17.1) is proved.

Here is the generalization to an arbitrary number of sets.

THEOREM 17.1 Inclusion–Exclusion Principle
Let n be a positive integer, and let $A_1, \ldots, A_n$ be finite sets. Then

$$|A_1 \cup \cdots \cup A_n| = c_1 - c_2 + c_3 - \cdots + (-1)^n c_n, \qquad (17.2)$$

where for $1 \leq i \leq n$, the number c_i is the sum of the sizes of the intersections of the sets taken i at a time.

In case any clarification is needed, for $n = 3$ we have

$$c_1 = |A_1| + |A_2| + |A_3|,$$
$$c_2 = |A_1 \cap A_2| + |A_1 \cap A_3| + |A_2 \cap A_3|,$$
$$c_3 = |A_1 \cap A_2 \cap A_3|,$$

so the theorem agrees with the equation (17.1).

PROOF   The argument is similar to the one I gave for the case $n = 3$ before stating the theorem. Let x be a member of the union $A_1 \cup \cdots \cup A_n$. We will show that x contributes exactly 1 to the right-hand side of the equation (17.2).

Suppose x belongs to precisely k of the sets $A_1, \ldots, A_n$. Then x contributes k to the sum $c_1 = |A_1| + \cdots + |A_n|$. In the sum $c_2 = \sum_{i<j} |A_i \cap A_j|$, x contributes 1 to all terms $|A_i \cap A_j|$ for which A_i and A_j are among the k sets containing x; there are $\binom{k}{2}$ such terms, so this is the contribution of x to the sum c_2. Similarly, x contributes $\binom{k}{3}$ to the sum c_3, and in general contributes $\binom{k}{i}$ to the sum c_i. Therefore the total contribution of x to the right-hand side of (17.2) is

$$k - \binom{k}{2} + \binom{k}{3} - \cdots + (-1)^{k-1} \binom{k}{k}.$$

By the equality (16.1) in the previous chapter, this is equal to 1. Hence each element $x \in A_1 \cup \cdots \cup A_n$ contributes exactly 1 to the right-hand side of (17.2), and the proof is complete.   ∎

Example 17.4
How many integers between 1 and 420 are divisible by 2, 3, 5 or 7?

Answer Notice that $420 = 2^2 \cdot 3 \cdot 5 \cdot 7$. Let A_2 be the set of integers between 1 and 420 that are divisible by 2, and define A_3, A_5, A_7 similarly. The question is asking for the size of the union $A_2 \cup A_3 \cup A_5 \cup A_7$. To apply the Inclusion–Exclusion Principle, we need to work out the sizes of the sets A_i, $A_i \cap A_j$, $A_i \cap A_j \cap A_k$ and $A_i \cap A_j \cap A_k \cap A_l$ for distinct $i, j, k, l \in \{2, 3, 5, 7\}$. This is straightforward: for example, $|A_2|$ is the number of multiplies of 2 that are between 1 and 420, which is $420/2 = 210$; and $|A_2 \cap A_3|$ is the number of multiples of 2×3, which is $420/6 = 70$; and so on. Hence using the Inclusion–Exclusion Principle, we see that $|A_2 \cup A_3 \cup A_5 \cup A_7| = c_1 - c_2 + c_3 - c_4$ where

$$
\begin{aligned}
c_1 &= |A_2| + \cdots + |A_7| = 210 + 140 + 84 + 60, \\
c_2 &= |A_2 \cap A_3| + \cdots + |A_5 \cap A_7| = 70 + 42 + 30 + 28 + 20 + 12, \\
c_3 &= |A_2 \cap A_3 \cap A_5| + \cdots + |A_3 \cap A_5 \cap A_7| = 14 + 10 + 6 + 4, \\
c_4 &= |A_2 \cap A_3 \cap A_5 \cap A_7| = 2.
\end{aligned}
$$

Therefore $|A_2 \cup A_3 \cup A_5 \cup A_7| = 324$.

The Inclusion–Exclusion Principle has many applications. Next we will give a nice one to the theory of numbers.

DEFINITION   *For a positive integer n, define $\phi(n)$ to be the number of integers x such that $1 \leq x \leq n$ and $\mathrm{hcf}(x, n) = 1$. The function ϕ is known as the* Euler ϕ-function.

For example, the set of integers between 1 and 10 that are coprime to 10 is $\{1, 3, 7, 9\}$, so $\phi(10) = 4$. Also Example 17.4 shows that $\phi(420) = 420 - 324 = 96$.

The next result gives a famous explicit formula for the Euler ϕ-function.

PROPOSITION 17.3
Let $n \geq 2$ be an integer with prime factorization $n = p_1^{a_1} p_2^{a_2} \cdots p_k^{a_k}$ (where the primes p_i are distinct and all $a_i \geq 1$). Then

$$
\phi(n) = n \left(1 - \frac{1}{p_1}\right) \left(1 - \frac{1}{p_2}\right) \cdots \left(1 - \frac{1}{p_k}\right).
$$

For example, $\phi(420) = 420 \cdot (1 - \frac{1}{2})(1 - \frac{1}{3})(1 - \frac{1}{5})(1 - \frac{1}{7}) = 96$, agreeing with Example 17.4.

PROOF   Let $I_n = \{1, \ldots, n\}$. For each $i \in \{1, \ldots, k\}$ define A_i to be the set of integers in I_n that are divisible by the prime p_i. Then

$$
\phi(n) = n - |A_1 \cup \cdots \cup A_k|. \tag{17.3}
$$

We work out the size of $A_1 \cup \cdots \cup A_k$ in similar fashion to Example 17.4. By the Inclusion–Exclusion Principle,

$$|A_1 \cup \cdots \cup A_k| = c_1 - c_2 + \cdots + (-1)^{k-1} c_k, \tag{17.4}$$

where c_i is the sum of the intersections of the sets taken i at a time. Consider an intersection $A_{j_1} \cap \cdots \cap A_{j_i}$. This consists of the multiples of $p_{j_1} \cdots p_{j_i}$ in I_n, of which there are $n / p_{j_1} \cdots p_{j_i}$. Consequently

$$c_1 = n \left(\frac{1}{p_1} + \frac{1}{p_2} + \cdots + \frac{1}{p_k} \right),$$

$$c_2 = n \left(\frac{1}{p_1 p_2} + \frac{1}{p_1 p_3} + \cdots + \frac{1}{p_{k-1} p_k} \right),$$

and so on. Hence by (17.3) and (17.4), $\phi(n)$ is equal to

$$n \left(1 - \left(\frac{1}{p_1} + \cdots + \frac{1}{p_k} \right) + \left(\frac{1}{p_1 p_2} + \cdots + \frac{1}{p_{k-1} p_k} \right) - \cdots + (-1)^k \frac{1}{p_1 \cdots p_k} \right).$$

This is equal to

$$n \left(1 - \frac{1}{p_1} \right) \left(1 - \frac{1}{p_2} \right) \cdots \left(1 - \frac{1}{p_k} \right),$$

which gives the result. ∎

To conclude the chapter, here is a neat and useful result about the number of subsets of a finite set.

PROPOSITION 17.4
Let S be a finite set consisting of n elements. Then the total number of subsets of S is equal to 2^n.

PROOF Let $S = \{1, 2, \ldots, n\}$. A subset $\{i_1, \ldots, i_k\}$ of S corresponds to a string consisting of n 0's and 1's, where we put 1's in positions $i_1, \ldots, i_k$ and 0's elsewhere; and every such string corresponds to a subset. (For example, if $n = 6$, the subset $\{2, 3, 5\}$ corresponds to the string 011010, and the string 100101 corresponds to the subset $\{1, 4, 6\}$.) Hence the number of subsets of S is equal to the number of strings consisting of n 0's and 1's. Since we have 2 choices (0 or 1) for each of the n entries of such a string, the number of strings is equal to 2^n. ∎

You are asked to provide two alternative proofs of this result in Exercise 9.

Exercises for Chapter 17

1. (a) Let A, B be sets. Prove that $A \cup B = A$ if and only if $B \subseteq A$.

    (b) Prove that $(A - C) \cap (B - C) = (A \cap B) - C$ for all sets A, B, C.

2. Which of the following statements are true and which are false? Give proofs or counterexamples.

    (a) For any sets A, B, C, we have

    $$A \cup (B \cap C) = (A \cup B) \cap (A \cup C).$$

    (b) For any sets A, B, C, we have

    $$(A - B) - C = A - (B - C).$$

    (c) For any sets A, B, C, we have

    $$(A - B) \cup (B - C) \cup (C - A) = A \cup B \cup C.$$

3. Work out $\bigcup_{n=1}^{\infty} A_n$ and $\bigcap_{n=1}^{\infty} A_n$, where A_n is defined as follows for $n \in \mathbb{N}$:

    (a) $A_n = \{x \in \mathbb{R} \mid x > n\}$.

    (b) $A_n = \{x \in \mathbb{R} \mid \frac{1}{n} < x < \sqrt{2} + \frac{1}{n}\}$.

    (c) $A_n = \{x \in \mathbb{R} \mid -n < x < \frac{1}{n}\}$.

    (d) $A_n = \{x \in \mathbb{Q} \mid \sqrt{2} - \frac{1}{n} \leq x \leq \sqrt{2} + \frac{1}{n}\}$.

4. (a) 73% of British people like cheese, 76% like apples and 10% like neither. What percentage like both cheese and apples?

    (b) In a class of 30 children, everyone supports at least one of three teams: 16 support Manchester United, 17 support Stoke City and 14 support Doncaster Rovers; also 8 support both United and City, 7 both United and Rovers, and 9 both City and Rovers. How many support all three teams?

5. How many integers are there between 1000 and 9999 that contain the digits 0, 8 and 9 at least once each? (For example, 8950 and 8089 are such integers.)

6. How many integers between 1 and 10000 are neither squares nor cubes?

7. How many integers between 2 and 10000 are r^{th} powers for some $r \in \{2, 3, 4, 5\}$?

8. (a) Find the number of integers between 1 and 5000 that are divisible by neither 3 nor 4.

(b) Find the number of integers between 1 and 5000 that are divisible by none of the numbers 3, 4 and 5.

(c) Find the number of integers between 1 and 5000 that are divisible by one or more of the numbers 4, 5 and 6.

9. (a) The equality

$$\sum_{r=0}^{n} \binom{n}{r} = 2^n$$

is given just after Proposition 16.3. Use this to give an alternative proof of Proposition 17.4.

(b) Give yet another proof of Proposition 17.4 by induction on n.

10. (a) Calculate $\phi(1000)$ and $\phi(999)$, where ϕ is Euler's ϕ-function.

(b) Find the minimum and maximum values of $\phi(n)$ for $20 \le n \le 30$.

(c) Show that if $n \ge 3$ then $\phi(n)$ is even.

(d) Find all positive integers n such that $\phi(n)$ is not divisible by 4.

11. Prove that if m and n are coprime positive integers, then $\phi(mn) = \phi(m)\phi(n)$.

12. For a positive integer n, define

$$F(n) = \sum_{d \mid n} \phi(d),$$

where the sum is over the positive divisors d of n, including both 1 and n. (For example, the positive divisors of 15 are 1, 3, 5 and 15.)

(a) Calculate $F(15)$ and $F(100)$.

(b) Calculate $F(p^r)$, where p is prime.

(c) Calculate $F(pq)$, where p, q are distinct primes.

(d) Formulate a conjecture about $F(n)$ for an arbitrary positive integer n. Try to prove your conjecture.

13. Let n be a positive integer, and let $D(n)$ be the set of arrangements of $\{1, \ldots, n\}$ for which no number is in its corresponding position. (For example, if $n = 4$ then the arrangement $4, 2, 3, 1$ is not in $D(4)$ as the number 2 is in position 2; but the arrangement $4, 3, 2, 1$ *is* in $D(4)$.) Use the Inclusion–Exclusion Principle to prove that

$$|D(n)| = n! \left(1 - \frac{1}{1!} + \frac{1}{2!} - \frac{1}{3!} + \cdots + (-1)^n \frac{1}{n!} \right).$$

14. Some time ago, critic Ivor Smallbrain threw a lavish party for 5 of his best friends, including a private showing of the fabulous new Ally Wooden film *Everything You Wanted to Know About Sets But Were Afraid to Ask*. It was a rainy day, and each guest brought their own umbrella. At the end of the party, somewhat the worse for wear, Ivor handed each friend a random umbrella, and off they went into the night.

Work out the chance that nobody got their own umbrella.

Do the same calculation for the three subsequent parties thrown by Ivor, to which he invited 6, 7 and 8 guests, respectively. Do you notice anything interesting?

Chapter 18

Equivalence Relations

Let S be a set. A *relation* on S is defined as follows. We choose a subset R of the Cartesian product $S \times S$; in other words, R consists of some of the ordered pairs (s,t) with $s,t \in S$. For those ordered pairs $(s,t) \in R$, we write $s \sim t$ and say s is *related* to t. And for $(s,t) \notin R$, we write $s \nsim t$. Thus, the symbol $\sim$ relates various pairs of elements of S. It is called a *relation* on S.

This definition probably seems a bit strange at first sight. A few examples should serve to clarify matters.

Example 18.1

Here are eight examples of relations on various sets S.

(1) Let $S = \mathbb{R}$, and define $a \sim b \Leftrightarrow a < b$. Here $R = \{(s,t) \in \mathbb{R} \times \mathbb{R} \mid s < t\}$.

(2) Let $S = \mathbb{Z}$ and let m be a positive integer. Define $a \sim b \Leftrightarrow a \equiv b \bmod m$.

(3) $S = \mathbb{C}$, and $a \sim b \Leftrightarrow |a - b| < 1$.

(4) $S = \mathbb{R}$, and $a \sim b \Leftrightarrow a + b \in \mathbb{Z}$.

(5) $S = \{1,2\}$, and $\sim$ defined by $1 \sim 1, 1 \sim 2, 2 \nsim 1, 2 \sim 2$.

(6) $S = \{1,2\}$, and $\sim$ defined by $1 \sim 1, 1 \nsim 2, 2 \nsim 1, 2 \sim 2$.

(7) $S = $ all people in Britain, and $a \sim b$ if and only if a and b have the same father.

(8) S any set, and $a \sim b \Leftrightarrow a = b$.

The relations on a set S correspond to the subsets of $S \times S$, and there is nothing much more to say about them in general. However, there are certain types of relations that are worthy of study, as they crop up frequently. These are called *equivalence relations*. Here is the definition.

DEFINITION *Let S be a set, and let $\sim$ be a relation on S. Then $\sim$ is an* equivalence relation *if the following three properties hold for all*

$a, b, c \in S$:

(i) $a \sim a$ (this says $\sim$ is reflexive)
(ii) if $a \sim b$ then $b \sim a$ (this says $\sim$ is symmetric)
(iii) if $a \sim b$ and $b \sim c$ then $a \sim c$ (this says $\sim$ is transitive)

Let us examine each of the Examples 18.1 for these properties.

In Example 18.1(1), $S = \mathbb{R}$ and $a \sim b \Leftrightarrow a < b$. This is not reflexive or symmetric, but it is transitive as $(a < b$ and $b < c) \Rightarrow a < c$.

Example 18.1(2) is an equivalence relation by Proposition 13.2.

Now consider Example 18.1(3), where $S = \mathbb{C}$ and $a \sim b \Leftrightarrow |a - b| < 1$. This is reflexive and symmetric. But it is not transitive; to see this, take $a = \frac{3}{4}, b = 0, c = -\frac{3}{4}$: then $|a - b| < 1, |b - c| < 1$ but $|a - c| > 1$.

The relation in Example 18.1(4) is symmetric but not reflexive or transitive; the relation in Example 18.1(5) is reflexive and transitive, but not symmetric; and the relation in Example 18.1(6) is an equivalence relation.

I leave it to you to show that the relations in Examples 18.1(7) and (8) are both equivalence relations.

Equivalence Classes

Let S be a set and $\sim$ an equivalence relation on S. For $a \in S$, define

$$cl(a) = \{s \,|\, s \in S, s \sim a\}.$$

Thus, $cl(a)$ is the set of things that are related to a. The subset $cl(a)$ is called an *equivalence class* of $\sim$. The equivalence classes of $\sim$ are the subsets $cl(a)$ as a ranges over the elements of S.

Example 18.2

Let m be a positive integer, and let $\sim$ be the equivalence relation on $\mathbb{Z}$ defined as in Example 18.1(2) — that is,

$$a \sim b \Leftrightarrow a \equiv b \bmod m.$$

What are the equivalence classes of this relation?

To answer this, let us write down various equivalence classes:

$$cl(0) = \{s \in \mathbb{Z} \,|\, s \equiv 0 \bmod m\},$$
$$cl(1) = \{s \in \mathbb{Z} \,|\, s \equiv 1 \bmod m\}, \ldots$$
$$cl(m-1) = \{s \in \mathbb{Z} \,|\, s \equiv m - 1 \bmod m\}.$$

We claim that these are *all* the equivalence classes. For if n is any integer, then by Proposition 10.1 there are integers q, r such that $n = qm + r$ with $0 \leq r < m$. Then $n \equiv r \bmod m$, so $n \in cl(r)$, which is one of the classes listed above, and moreover,

$$cl(n) = \{s \in \mathbb{Z} \mid s \equiv n \bmod m\} = \{s \in \mathbb{Z} \mid s \equiv r \bmod m\} = cl(r).$$

Hence, any equivalence class $cl(n)$ is equal to one of those listed above.

We conclude that in this example, there are exactly m different equivalence classes, $cl(0), cl(1), \ldots, cl(m-1)$. Note also that every integer lies in exactly one of these classes.

Example 18.3

Consider now the equivalence relation defined in Example 18.1(7): $S =$ all people in Britain, and $a \sim b$ if and only if a and b have the same father. What are the equivalence classes?

If $a \in S$, then $cl(a)$ is the set of all people with the same father as a. In other words, if f is the father of a, then $cl(a)$ consists of all the children of f. So one way of listing all the equivalence classes is as follows: let $f_1, \ldots, f_n$ be a list of all fathers of people in Britain; if C_i is the set of children of f_i living in Britain, then the equivalence classes are $C_1, \ldots, C_n$.

We now prove a general property of equivalence classes. Recall from Chapter 16 that a *partition* of a set S is a collection of subsets $S_1, \ldots, S_k$ such that each element of S lies in exactly one of these subsets. Another way of putting this is that the subsets $S_1, \ldots, S_k$ have the properties that their union is S and any two of them are disjoint (i.e., $S_i \cap S_j = \emptyset$ for any $i \neq j$).

For example, if $S = \{1, 2, 3, 4, 5\}$, then the subsets $\{1\}, \{2, 4\}, \{3, 5\}$ form a partition of S, whereas the subsets $\{1\}, \{2, 4\}, \{3\}, \{4, 5\}$ do not.

PROPOSITION 18.1

Let S be a set and let $\sim$ be an equivalence relation on S. Then the equivalence classes of $\sim$ form a partition of S.

PROOF   If $a \in S$, then since $a \sim a$, a lies in the equivalence class $cl(a)$.

We need to show that a lies in only one equivalence class. So suppose that a lies in $cl(s)$ and $cl(t)$; in other words, $a \sim s$ and $a \sim t$. We show that this implies that $cl(s) = cl(t)$.

Let $x \in cl(s)$. Then $x \sim s$. Also $s \sim a$ and $a \sim t$, so by transitivity, $x \sim t$. Hence

$$x \in cl(s) \Rightarrow x \in cl(t).$$

Similarly, if $x \in cl(t)$ then $x \sim t$, and also $t \sim a$ and $a \sim s$, so $x \sim s$, showing

$$x \in cl(t) \Rightarrow x \in cl(s).$$

We conclude that $cl(s) = cl(t)$, as required. Thus, any element a of S lies in exactly one equivalence class. ∎

To conclude, observe that Proposition 18.1 is true the other way round: if S is a set and $S_1, \ldots, S_k$ is a partition of S, then there is a unique equivalence relation $\sim$ on S which has the S_i as its equivalence classes — namely, the equivalence relation defined as follows: for $x, y \in S$,

$$x \sim y \Leftrightarrow \text{there exists } i \text{ such that } x, y \text{ both lie in } S_i.$$

You are asked to justify this statement in Exercise 4 at the end of the chapter. In case it does not strike you as being completely obvious, here's an example: if $S = \{1, 2, 3, 4\}$, then the equivalence relation corresponding to the partition $\{1, 3\}, \{2\}, \{4\}$ is $\sim$, where

$$1 \sim 1, 1 \sim 3, 3 \sim 1, 3 \sim 3, 2 \sim 2, 4 \sim 4,$$

and no other pairs are related.

The upshot of all this is that there is a very tight correspondence between the equivalence relations on a set S and the partitions of S: every equivalence relation gives a unique partition of S (namely, the collection of equivalence classes); and every partition gives a unique equivalence relation (namely, the relation defined above). This makes classifying equivalence relations pretty easy — for example, the number of equivalence relations on S is equal to the number of partitions of S.

Exercises for Chapter 18

1. Which of the following relations are equivalence relations on the given set S?

   (i) $S = \mathbb{R}$, and $a \sim b \Leftrightarrow a = b$ or $-b$.

   (ii) $S = \mathbb{Z}$, and $a \sim b \Leftrightarrow ab = 0$.

   (iii) $S = \mathbb{R}$, and $a \sim b \Leftrightarrow a^2 + a = b^2 + b$.

   (iv) S is the set of all people in the world, and $a \sim b$ means a lives within 100 miles of b.

(v) S is the set of all points in the plane, and $a \sim b$ means a and b are the same distance from the origin.

(vi) $S = \mathbb{N}$, and $a \sim b \Leftrightarrow ab$ is a square.

(vii) $S = \{1,2,3\}$, and $a \sim b \Leftrightarrow a = 1$ or $b = 1$.

(viii) $S = \mathbb{R} \times \mathbb{R}$, and $(x,y) \sim (a,b) \Leftrightarrow x^2 + y^2 = a^2 + b^2$.

2. For those relations in Exercise 1 that are equivalence relations, describe the equivalence classes.

3. By producing suitable examples of relations, show that it is not possible to deduce any one of the properties of being reflexive, symmetric or transitive from the other two.

4. Prove that if S is a set and $S_1, \ldots, S_k$ is a partition of S, then there is a unique equivalence relation $\sim$ on S that has the S_i as its equivalence classes.

5. (a) How many relations are there on the set $\{1,2\}$?

(b) How many relations are there on the set $\{1,2,3\}$ that are both reflexive and symmetric?

(c) How many relations are there on the set $\{1,2,\ldots,n\}$?

6. Let $S = \{1,2,3,4\}$, and suppose that $\sim$ is an equivalence relation on S. You are given the information that $1 \sim 2$ and $2 \sim 3$.

Show that there are exactly two possibilities for the relation $\sim$, and describe both (i.e., for all $a,b \in S$, say whether or not $a \sim b$).

7. Let $\sim$ be an equivalence relation on $\mathbb{Z}$ with the property that for all $m \in \mathbb{Z}$ we have $m \sim m+5$ and also $m \sim m+8$. Prove that $m \sim n$ for all $m,n \in \mathbb{Z}$.

8. Critic Ivor Smallbrain has made his peace with rival Greta Picture, and they are now friends. Possibly their friendship will develop into something even more beautiful, who knows. Ivor and Greta are sitting through a showing of the latest Disney film, *101 Equivalence Relations*. They are fed up and start to discuss how many different equivalence relations they can find on the set $\{1,2\}$. They find just two. Then on the set $\{1,2,3\}$ they find just five different equivalence relations.

Have they found *all* the equivalence relations on these sets? How many should they find on $\{1,2,3,4\}$ and on $\{1,2,3,4,5\}$? Investigate further if you feel like it!

Chapter 19

Functions

Much of mathematics and its applications is concerned with the study of functions of various kinds. In this chapter we give the definition and some elementary examples, and introduce certain important general types of functions.

DEFINITION   *Let S and T be sets. A function from S to T is a rule that assigns to each $s \in S$ a single element of T, denoted by $f(s)$. We write*

$$f : S \rightarrow T$$

to mean that f is a function from S to T. If $f(s) = t$, we often say f sends $s \rightarrow t$.

   If $f : S \rightarrow T$ is a function, the image of f is the set of all elements of T that are equal to $f(s)$ for some $s \in S$. We write $f(S)$ for the image of f. Thus

$$f(S) = \{f(s) \,|\, s \in S\}.$$

Example 19.1
(1) Define $f : \{1,2,3\} \rightarrow \mathbb{Z}$ by $f(x) = x^2 - 4$ for $x \in \{1,2,3\}$. The image of f is $\{-3,0,5\}$.

(2) Define $f : \mathbb{R} \rightarrow \mathbb{R}$ by $f(x) = x^2$ for all $x \in \mathbb{R}$. The image of f is $f(\mathbb{R}) = \{y \,|\, y \in \mathbb{R}, y \geq 0\}$.

(3) A body is dropped and falls under gravity for 1 second. The distance travelled at time t is $\frac{1}{2}gt^2$. If we call this distance $s(t)$ and write $I = \{t \in \mathbb{R} \,|\, 0 \leq t \leq 1\}$, then s is a function from I to $\mathbb{R}$ defined by $s(t) = \frac{1}{2}gt^2$. The image of s is the set of reals between 0 and $\frac{g}{2}$.

(4) Define $f : \mathbb{N} \times \mathbb{N} \rightarrow \mathbb{Z}$ by $f(m,n) = m - n$ for all $m,n \in \mathbb{N}$. The image of f is $\mathbb{Z}$.

(5) Let $S = \{a,b,c\}$ and define functions $f : S \rightarrow S$ and $g : S \rightarrow S$ as follows:

$$f \text{ sends } a \rightarrow b, \ b \rightarrow c, \ c \rightarrow a, \quad g \text{ sends } a \rightarrow b, \ b \rightarrow c, \ c \rightarrow b.$$

Then $f(S) = S$, while $g(S) = \{b, c\}$.

(6) Let S be any set, and define a function $\iota_S : S \to S$ by

$$\iota_S(s) = s \quad \text{for all } s \in S.$$

This function ι_S is called the *identity* function of S.

We now define certain important types of functions.

DEFINITION   *Let $f : S \to T$ be a function.*

(I) We say f is onto *if the image $f(S) = T$; i.e., if for every $t \in T$ there exists $s \in S$ such that $f(s) = t$.*

(II) We say f is one-to-one *(usually written simply as 1-1) if whenever $s_1, s_2 \in S$ with $s_1 \neq s_2$, then $f(s_1) \neq f(s_2)$; in other words, f is 1-1 if f sends different elements of S to different elements of T. Another way of putting this is to say that for all $s_1, s_2 \in S$,*

$$f(s_1) = f(s_2) \Rightarrow s_1 = s_2.$$

This is usually the most useful definition to use when testing whether functions are 1-1.

(III) We say f is a bijection *if f is both onto and 1-1.*

Functions that are onto are often called *surjective* functions, or *surjections*; and functions that are 1-1 are often called *injective* functions, or *injections*. You will find these terms in many books, but I prefer to stick to the slightly more descriptive terms "onto" and "1-1."

Let us briefly discuss which of these properties the functions in Example 19.1 possess.

The function in Example 19.1(1) sends $1 \to -3, 2 \to 0, 3 \to 5$, so it is 1-1. It is clearly not onto.

The function in Example 19.1(2) is not onto and is not 1-1 either, since it sends 1 and -1 to the same thing.

On the other hand, the function $s : I \to \mathbb{R}$ in Example 19.1(3) is 1-1, since, for $t_1, t_2 \in I$,

$$s(t_1) = s(t_2) \Rightarrow \frac{1}{2}g t_1^2 = \frac{1}{2}g t_2^2 \Rightarrow t_1 = t_2.$$

Also, s is not onto.

The function in Example 19.1(4) is onto but is not 1-1 since, for example, it sends both $(1, 1)$ and $(2, 2)$ to 0.

In Example 19.1(5), the function f is a bijection, while g is neither 1-1 nor onto. Finally, the identity function in Example 19.1(6) is a bijection.

Here is a quite useful result relating 1-1 and onto functions to the sizes of sets.

PROPOSITION 19.1

Let $f : S \to T$ be a function, where S and T are finite sets.

(i) *If f is onto, then $|S| \geq |T|$.*
(ii) *If f is 1-1, then $|S| \leq |T|$.*
(iii) *If f is a bijection, then $|S| = |T|$.*

PROOF (i) Let $|S| = n$ and write $S = \{s_1, \ldots, s_n\}$. As f is onto, we have

$$T = f(S) = \{f(s_1), \ldots, f(s_n)\}.$$

Hence $|T| \leq n$. (Of course $|T|$ could be less than n, as some of the $f(s_i)$'s could be equal.)

(ii) Again let $|S| = n$ and $S = \{s_1, \ldots, s_n\}$. As f is 1-1, the elements $f(s_1), \ldots, f(s_n)$ are all different and lie in T. Therefore $|T| \geq n$.

(iii) If f is a bijection, then $|S| \geq |T|$ by (i) and $|S| \leq |T|$ by (ii), so $|S| = |T|$. ∎

The Pigeonhole Principle

Part (ii) of Proposition 19.1 implies that if $|S| > |T|$, then there is no 1-1 function from S to T. This can be phrased somewhat more strikingly in the following way:

If we put $n + 1$ or more pigeons into n pigeonholes, then there must be a pigeonhole containing more than one pigeon.

(For if no pigeonhole contained more than one pigeon, the function sending pigeons to their pigeonholes would be 1-1.)

The above statement is known as the *Pigeonhole Principle*, and it is surprisingly useful. As a very simple example, in any group of 13 or more people, at least two must have their birthday in the same month (here the people are the "pigeons" and the 12 months are the "pigeonholes"). As another example, in any set of 6 integers, there must be two whose difference is divisible by 5: to see this, regard the 6 integers as the pigeons, and their remainders on division by 5 as the pigeonholes.

Here's a slightly more subtle example.

Example 19.2
Prove that if $n+1$ numbers are chosen from the set $\{1,2,\ldots,2n\}$, there will always be two of the chosen numbers that differ by 1. (This is not necessarily true if we choose only n numbers — for example, we could choose the n numbers $1,3,5,\ldots,2n-1$.)

Answer This becomes easy when we make the following cunning choice of what the pigeonholes are. Define the pigeonholes to be the n sets

$$\{1,2\},\{3,4\},\ldots,\{2n-1,2n\}.$$

Since we are choosing $n+1$ numbers, each of which belongs to one of these pigeonholes, the Pigeonhole Principle tells us that two of them must lie in the same pigeonhole. These two will then differ by 1. Pretty neat, eh?

More examples of the use of the Pigeonhole Principle can be found in Exercise 5 at the end of the chapter.

Inverse Functions

Given a function $f : S \to T$, under what circumstances can we define an "inverse function" from T to S, sending everything back to where it came from? (In other words, if f sends $s \to t$, the "inverse" function should send $t \to s$.) To define such a function from T to S, we need:

(a) f to be onto (otherwise some elements of T will not be sent anywhere by the inverse function), and

(b) f to be 1-1 (otherwise some element of T will be sent back to more than one element of S).

In other words, to be able to define such an inverse function from T to S, we need f to be a bijection. Here is the formal definition.

DEFINITION   *Let $f : S \to T$ be a bijection. The inverse function of f is the function from $T \to S$ that sends each $t \in T$ to the unique $s \in S$ such that $f(s) = t$. We denote the inverse function by $f^{-1} : T \to S$. Thus, for $s \in S, t \in T$,*

$$f^{-1}(t) = s \Leftrightarrow f(s) = t.$$

As a consequence we have

$$f^{-1}(f(s)) = s \quad and \quad f(f^{-1}(t)) = t$$

for all $s \in S, t \in T$.

Example 19.3

(1) Let $S = \{a,b,c\}$ and let $f : S \rightarrow S$ be the function that sends $a \rightarrow b, b \rightarrow c, c \rightarrow a$. Then f is a bijection and the inverse function $f^{-1} : S \rightarrow S$ sends everything back to where it came from; namely, $a \rightarrow c, b \rightarrow a, c \rightarrow b$.

(2) Define $f : \mathbb{R} \rightarrow \mathbb{R}$ by $f(x) = 8 - 2x$ for all $x \in \mathbb{R}$. Then f is a bijection and $f^{-1}(t) = \frac{1}{2}(8 - t)$ for all $t \in \mathbb{R}$.

Composition of Functions

Composition gives us a useful way of combining two functions to form another one. Here is the definition.

DEFINITION   *Let S,T,U be sets, and let $f : S \rightarrow T$ and $g : T \rightarrow U$ be functions. The* composition *of f and g is the function $g \circ f : S \rightarrow U$, which is defined by the rule*

$$(g \circ f)(s) = g(f(s)) \quad \text{for all } s \in S.$$

Thus $g \circ f$ is just a "function of a function," which is a phrase you may have seen before.

Example 19.4

(1) Let $f : \mathbb{R} \rightarrow \mathbb{R}, \; g : \mathbb{R} \rightarrow \mathbb{R}$ be defined by

$$f(x) = \sin x, \; g(x) = x^2 + 1$$

for all $x \in \mathbb{R}$. Then both compositions $g \circ f$ and $f \circ g$ are functions from $\mathbb{R} \rightarrow \mathbb{R}$, and

$$(g \circ f)(x) = g(f(x)) = \sin^2 x + 1, \quad f \circ g(x) = \sin\left(x^2 + 1\right)$$

for all $x \in \mathbb{R}$.

(2) Let $f : \{1,2,3\} \rightarrow \mathbb{Z}$ and $g : \mathbb{Z} \rightarrow \mathbb{N}$ be defined by

$$f \text{ sends } 1 \rightarrow 0, 2 \rightarrow -5, 3 \rightarrow 7 \quad \text{and}$$
$$g(x) = |x| \quad \text{for all } x \in \mathbb{Z}.$$

Then $g \circ f : \{1,2,3\} \rightarrow \mathbb{N}$ sends $1 \rightarrow 0, 2 \rightarrow 5, 3 \rightarrow 7$ and $f \circ g$ does not exist.

Notice that if $f : S \to T$ is a bijection, then by definition of the inverse function $f^{-1} : T \to S$, we have

$$\left(f^{-1} \circ f\right)(s) = s, \quad \left(f \circ f^{-1}\right)(t) = t$$

for all $s \in S, t \in T$. Another way of putting this is to say that

$$f^{-1} \circ f = \iota_S, \quad f \circ f^{-1} = \iota_T,$$

where ι_S, ι_T are the identity functions of S and T, as defined in Example 19.1(6).

Here is a neat result linking composition with the properties of being 1-1 or onto.

PROPOSITION 19.2
Let S, T, U be sets, and let $f : S \to T$ and $g : T \to U$ be functions. Then

(i) *if f and g are both 1-1, so is $g \circ f$,*
(ii) *if f and g are both onto, so is $g \circ f$,*
(iii) *if f and g are both bijections, so is $g \circ f$.*

PROOF   (i) If f, g are both 1-1, then for $s_1, s_2 \in S$,

$$
\begin{aligned}
(g \circ f)(s_1) = (g \circ f)(s_2) &\Rightarrow g(f(s_1)) = g(f(s_2)) \\
&\Rightarrow f(s_1) = f(s_2) \text{ as } g \text{ is 1-1} \\
&\Rightarrow s_1 = s_2 \text{ as } f \text{ is 1-1}
\end{aligned}
$$

and hence $g \circ f$ is 1-1.

(ii) Suppose f, g are both onto. For any $u \in U$, there exists $t \in T$ such that $g(t) = u$ (as g is onto), and there exists $s \in S$ such that $f(s) = t$ (as f is onto). Hence $(g \circ f)(s) = g(f(s)) = g(t) = u$, showing that $g \circ f$ is onto.

(iii) This follows immediately from parts (i) and (ii).   ∎

Counting Functions

How many functions are there from one finite set to another? This question is quite easily answered using some of our counting methods from Chapter 16.

PROPOSITION 19.3
Let S, T be finite sets, with $|S| = m$, $|T| = n$. Then the number of functions from S to T is equal to n^m.

PROOF Let $S = \{s_1, s_2, \ldots, s_m\}$. Defining a function $f : S \to T$ is an m-stage process:

Stage 1: Choose $f(s_1)$; this can be any of the n members of T, so the number of choices is n.

Stage 2: Choose $f(s_2)$; again, the number of choices is n.

And so on, up to

Stage m: Choose $f(s_m)$; again, the number of choices is n.

Thus, by the Multiplication Principle 16.1, the total number of functions is $n.n. \ldots n = n^m$. ∎

One can also obtain a formula for the number of 1-1 functions from S to T (see Exercise 7 at the end of this chapter). Counting onto functions is somewhat harder, and will have to wait until Exercise 7 at the end of the next chapter.

Exercises for Chapter 19

1. For each of the following functions f, say whether f is 1-1 and whether f is onto:

    (i) $f : \mathbb{R} \to \mathbb{R}$ defined by $f(x) = x^2 + 2x$ for all $x \in \mathbb{R}$.

    (ii) $f : \mathbb{R} \to \mathbb{R}$ defined by

    $$f(x) = \begin{cases} x - 2, & \text{if } x > 1 \\ -x, & \text{if } -1 \leq x \leq 1 \\ x + 2, & \text{if } x < -1. \end{cases}$$

    (iii) $f : \mathbb{Q} \to \mathbb{R}$ defined by $f(x) = (x + \sqrt{2})^2$.

    (iv) $f : \mathbb{N} \times \mathbb{N} \times \mathbb{N} \to \mathbb{N}$ defined by $f(m, n, r) = 2^m 3^n 5^r$ for all $m, n, r \in \mathbb{N}$.

    (v) $f : \mathbb{N} \times \mathbb{N} \times \mathbb{N} \to \mathbb{N}$ defined by $f(m, n, r) = 2^m 3^n 6^r$ for all $m, n, r \in \mathbb{N}$.

    (vi) Let $\sim$ be the equivalence relation on $\mathbb{Z}$ defined by $a \sim b \Leftrightarrow a \equiv b \bmod 7$, and let S be the set of equivalence classes of $\sim$. Define $f : S \to S$ by $f(\text{cl}(s)) = \text{cl}(s + 1)$ for all $s \in \mathbb{Z}$.

2. The functions $f, g : \mathbb{R} \to \mathbb{R}$ are defined as follows:

    $$f(x) = 2x \text{ if } 0 \leq x \leq 1, \text{ and } f(x) = 1 \text{ otherwise};$$
    $$g(x) = x^2 \text{ if } 0 \leq x \leq 1, \text{ and } g(x) = 0 \text{ otherwise}.$$

Give formulae describing the functions $g \circ f$ and $f \circ g$. Draw the graphs of these functions.

3. Two functions $f,g : \mathbb{R} \to \mathbb{R}$ are such that for all $x \in \mathbb{R}$,

$$g(x) = x^2 + x + 3, \quad \text{and} \quad (g \circ f)(x) = x^2 - 3x + 5.$$

Find the possibilities for f.

4. Let X, Y, Z be sets, and let $f : X \to Y$ and $g : Y \to Z$ be functions.

(a) Given that $g \circ f$ is onto, can you deduce that f is onto? Give a proof or a counterexample.

(b) Given that $g \circ f$ is onto, can you deduce that g is onto?

(c) Given that $g \circ f$ is 1-1, can you deduce that f is 1-1?

(d) Given that $g \circ f$ is 1-1, can you deduce that g is 1-1?

5. Use the Pigeonhole Principle to prove the following statements involving a positive integer n:

(a) In any set of 6 integers, there must be two whose difference is divisible by 5.

(b) In any set of $n+1$ integers, there must be two whose difference is divisible by n.

(c) Given any n integers $a_1, a_2, \ldots, a_n$, there is a non-empty subset of these whose sum is divisible by n. (*Hint:* Consider the integers $0, a_1$, $a_1 + a_2, \ldots, a_1 + \cdots + a_n$ and use (b).)

(d) Given any set S consisting of ten distinct integers between 1 and 50, there are two different 5-element subsets of S with the same sum.

(e) Given any set T consisting of nine distinct integers between 1 and 50, there are two disjoint subsets of T with the same sum.

(f) In any set of 101 integers chosen from the set $\{1, 2, \ldots, 200\}$, there must be two integers such that one divides the other.

6. (a) Find an onto function from $\mathbb{N}$ to $\mathbb{Z}$.

(b) Find a 1-1 function from $\mathbb{Z}$ to $\mathbb{N}$.

7. (a) Let $S = \{1,2,3\}$ and $T = \{1,2,3,4,5\}$. How many functions are there from S to T ? How many of these are 1-1?

(b) Let $|S| = m, |T| = n$ with $m \le n$. Show that the number of 1-1 functions from S to T is equal to $n(n-1)(n-2)\cdots(n-m+1)$.

8. The manufacturers of the high-fibre cereal "Improve Your Functions" are offering a prize of £1000 to anyone who can find three different integers a, b, c and a polynomial $P(x)$ with integer coefficients, such that

$$P(a) = b, \ P(b) = c \ \text{ and } P(c) = a.$$

Critics Ivor Smallbrain and Greta Picture spend several long evenings trying to solve this, without success.

Prove that nobody will win the prize.

(*Hint:* Observe that $P(x) - P(y) = (x - y)Q(x,y)$, where $Q(x,y)$ is a polynomial in x, y with integer coefficients. Substitute $x = a, y = b$, etc., into this equation and see what happens.)

Chapter 20

Permutations

Let S be a set. By a *permutation* of S, we mean a bijection from S to S — that is, a function from S to S that is both onto and 1-1. Permutations form a rather pleasant and useful class of functions, and we shall study various aspects of them in this chapter.

Example 20.1

(1) Let $S = \{1,2,3,4,5\}$ and let $f,g : S \to S$ be defined as follows:

$$f : 1 \to 2, 2 \to 4, 3 \to 3, 4 \to 5, 5 \to 1,$$

$$g : 1 \to 3, 2 \to 4, 3 \to 1, 4 \to 2, 5 \to 4.$$

Then f is a permutation of S, but g is not.

(2) The function $f : \mathbb{R} \to \mathbb{R}$, defined by $f(x) = 8 - 2x$ for all $x \in \mathbb{R}$, is a permutation of $\mathbb{R}$ [see Example 19.3(2)].

Frequently, we consider permutations of the set $\{1,2,\ldots,n\}$. Denote by S_n the set of all permutations of $\{1,2,\ldots,n\}$. For a permutation $f \in S_n$, we use the notation

$$\begin{pmatrix} 1 & 2 & \cdots & n \\ f(1) & f(2) & \cdots & f(n) \end{pmatrix}$$

to describe f. This notation completely specifies what f is, since it gives the value of $f(i)$ for every $i \in \{1,\ldots,n\}$. For example, the permutation f of Example 20.1(1) is

$$\begin{pmatrix} 1 & 2 & 3 & 4 & 5 \\ 2 & 4 & 3 & 5 & 1 \end{pmatrix}.$$

PROPOSITION 20.1

The number of permutations in S_n is $n!$.

PROOF   If $f \in S_n$, then the sequence $f(1), \ldots, f(n)$ is just an arrangement of the numbers $1, \ldots, n$ in some order. Hence the number of permutations in S_n is equal to the number of such arrangements, which is $n!$, by Proposition 16.1.   ∎

Example 20.2
There are $3! = 6$ permutations in S_3. Here they are:

$$\iota = \begin{pmatrix} 1 & 2 & 3 \\ 1 & 2 & 3 \end{pmatrix}, \quad f_2 = \begin{pmatrix} 1 & 2 & 3 \\ 1 & 3 & 2 \end{pmatrix}, \quad f_3 = \begin{pmatrix} 1 & 2 & 3 \\ 3 & 2 & 1 \end{pmatrix},$$

$$f_4 = \begin{pmatrix} 1 & 2 & 3 \\ 2 & 1 & 3 \end{pmatrix}, \quad f_5 = \begin{pmatrix} 1 & 2 & 3 \\ 2 & 3 & 1 \end{pmatrix}, \quad f_6 = \begin{pmatrix} 1 & 2 & 3 \\ 3 & 1 & 2 \end{pmatrix}.$$

Notice that ι is the function that sends everything to itself. We called this the identity function in Example 19.1(6).

For any set S, the identity function ι_S is a permutation of S, which we call the *identity* permutation. We shall usually just write it as ι.

Composition of Permutations

If f and g are permutations of a set S, the composition $f \circ g$ is defined, as in the previous chapter, by

$$f \circ g(s) = f(g(s)) \quad \text{for all } s \in S,$$

and by Proposition 19.2(iii), $f \circ g$ is also a permutation of S. In dealing with permutations we usually drop the "∘" symbol and write just fg, instead of $f \circ g$.

Example 20.3

(1) If f_2, f_3 are the permutations in S_3 given in Example 20.2, then $f_2 f_3$ sends $1 \to 2, 2 \to 3, 3 \to 1$ (remember $f_2 f_3$ means "first do f_3, then do f_2"), and so

$$f_2 f_3 = \begin{pmatrix} 1 & 2 & 3 \\ 2 & 3 & 1 \end{pmatrix} = f_5.$$

Similarly, $f_3 f_2$ sends $1 \to 3, 2 \to 1, 3 \to 2$, so $f_3 f_2 = f_6$. Notice that $f_2 f_3 \neq f_3 f_2$. So the order in which we form the composition of two permutations is very important.

(2) If we felt like it, we could form the composition of any two of the permutations in S_3 and put the answers in a kind of "multiplication table" for S_3. Here is the top left-hand corner of the table. You are asked to fill in the rest in Exercise 1 at the end of the chapter.

$$
\begin{array}{c|cccc}
 & \iota & f_2 & f_3 & \cdots \\
\hline
\iota & \iota & f_2 & f_3 & \cdots \\
f_2 & f_2 & \iota & f_5 & \cdots \\
f_3 & f_3 & f_6 & \iota & \cdots \\
\cdots & & & &
\end{array}
$$

We often refer to the composition fg of two permutations as the *product* of f and g; likewise, we speak of *multiplying* f and g to form their product.

Composition also allows us to define *powers* of permutations in a natural way. If f is a permutation of a set S, define f^2 to be the permutation $ff = f \circ f$ (i.e., the permutation obtained by "doing f twice"). Then define f^3 to be $f^2 f$, then $f^4 = f^3 f$, and so on. For example, if

$$
f = \begin{pmatrix} 1 & 2 & 3 & 4 & 5 \\ 2 & 4 & 3 & 5 & 1 \end{pmatrix},
$$

then

$$
f^2 = \begin{pmatrix} 1 & 2 & 3 & 4 & 5 \\ 4 & 5 & 3 & 1 & 2 \end{pmatrix}, \quad f^3 = \begin{pmatrix} 1 & 2 & 3 & 4 & 5 \\ 5 & 1 & 3 & 2 & 4 \end{pmatrix}.
$$

The *inverse* of a permutation f of a set S was defined in the previous chapter as the function f^{-1} sending everything back to where it came from. So f^{-1} is also a permutation and

$$
ff^{-1} = f^{-1} f = \iota,
$$

where ι is the identity permutation of S. For example, if $f \in S_5$ is as above, then

$$
f^{-1} = \begin{pmatrix} 1 & 2 & 3 & 4 & 5 \\ 5 & 1 & 3 & 2 & 4 \end{pmatrix}.
$$

Four Fundamental Features

There are four properties of composition of permutations that are of basic importance, and we list them in the next result. Actually we have already seen some of these properties, but they are worth highlighting anyway.

PROPOSITION 20.2
The following properties are true for the set S_n of all permutations of $\{1, 2, \ldots, n\}$.
   (i) If f and g are in S_n, so is fg.
   (ii) For any $f, g, h \in S_n$,

$$f(gh) = (fg)h.$$

   (iii) The identity permutation $\iota \in S_n$ satisfies

$$f\iota = \iota f = f$$

for any $f \in S_n$.
   (iv) Every permutation $f \in S_n$ has an inverse $f^{-1} \in S_n$ such that

$$ff^{-1} = f^{-1}f = \iota.$$

PROOF   We have already seen properties (i) and (iv), and (iii) is easy, since $f\iota(s) = f(\iota(s)) = f(s)$ for all $s \in S$, hence $f\iota = f$ and similarly $\iota f = f$.

Property (ii) is a little more subtle (but only a little): for $s \in S$,

$$(f(gh))(s) = f((gh)(s)) = f(g(h(s))),$$

while

$$((fg)h)(s) = (fg)(h(s)) = f(g(h(s))).$$

In other words, both $f(gh)$ and $(fg)h$ are the function "first do h, then do g, then do f." Hence $f(gh) = (fg)h$.   ∎

Property (ii) is known as "associativity" and means that we can multiply several permutations without worrying about how we bracket them. I gave you a glimpse of how crucial associativity can be in Chapter 2 (see Rules 2.1 and the discussion afterwards).

The four properties in Proposition 20.2 are the four axioms of what is known as "group theory" and tell us that S_n, together with the rule of composition, is a "group" (it is known as the *symmetric group of degree n*). I shall introduce you formally to group theory in Chapters 25 and 26.

The Cycle Notation

The notation

$$\begin{pmatrix} 1 & 2 & \cdots & n \\ f(1) & f(2) & \cdots & f(n) \end{pmatrix}$$

for permutations is rather cumbersome and is not particularly convenient for performing calculations. There is a much more compact and useful notation — the *cycle notation* for permutations — which I will now describe.

Consider the following permutation in S_8:

$$f = \begin{pmatrix} 1 & 2 & 3 & 4 & 5 & 6 & 7 & 8 \\ 4 & 5 & 6 & 3 & 2 & 7 & 1 & 8 \end{pmatrix}.$$

This sends 1 to 4, 4 to 3, 3 to 6, 6 to 7, and 7 back to 1; we say that the symbols 1,4,3,6,7 form a *cycle* of f (of length 5). Similarly, 2 and 5 form a cycle of length 2, and 8 forms a cycle of length 1. We write

$$f = (14367)(25)(8).$$

This notation indicates that each number 1,4,3,6,7 in the first cycle goes to the next one, except for the last, which goes back to the first; and likewise for the second and third cycles. This is the cycle notation for f. Notice that the cycles have no symbols in common; they are called *disjoint* cycles.

It is easy to move back from the cycle notation to the original notation, if desired. For example, if $g = (1372)(46)(5) \in S_7$, then the original notation for g is

$$g = \begin{pmatrix} 1 & 2 & 3 & 4 & 5 & 6 & 7 \\ 3 & 1 & 7 & 6 & 5 & 4 & 2 \end{pmatrix}.$$

It is not too hard to generalize this idea to arbitrary permutations.

DEFINITION   *For a set $S = \{a_1, \ldots, a_r\}$, the cycle $(a_1 a_2 \ldots a_r)$ is the permutation of S that sends $a_1 \to a_2, a_2 \to a_3, \ldots a_{r-1} \to a_r$ and $a_r \to a_1$. The* length *of the cycle is r, and we also call it an r-cycle.*

A collection of cycles is disjoint *if no two of the cycles have a symbol in common.*

PROPOSITION 20.3
Every permutation in S_n can be expressed as a product of disjoint cycles.

PROOF   Let $f \in S_n$. Begin with 1, and write down the sequence $1, f(1), f^2(1), f^3(1), \ldots$. These elements all lie in the finite set $\{1, \ldots, n\}$, so they can't all be distinct; let $f^r(1)$ be the first element in the sequence that has appeared previously. Then $f^r(1) = 1$: because if not, $f^r(1) = f^s(1)$ with $0 < s < r$, and then $f^{-s} f^r(1) = f^{-s} f^s(1)$, so $f^{r-s}(1) = 1$ with $r - s < r$, contradicting our choice of r. Now the cycle

$$c_1 = (1 f(1) f^2(1) \ldots f^{r-1}(1))$$

has the same effect as f on all the symbols in the cycle; this is the first cycle of f. To get the second cycle, choose (if possible) a symbol not appearing in c_1 — call it i. Then repeat the above process: write down the sequence $i, f(i), f^2(i), \ldots$ until we reach i again. Say $f^s(i) = i$; then $c_2 = (i\, f(i) \ldots f^{s-1}(i))$ is the second cycle of f. Note that c_2 and c_1 have no symbols in common, for if they had a symbol j in common, they would be identical, since each cycle could be constructed by repeatedly applying the permutation f starting at j.

Carrying on, we choose (if possible) a symbol not appearing in c_1 or c_2, and construct a cycle c_3, and so on. Since $\{1, \ldots, n\}$ is a finite set, this process must terminate with some cycle c_m. Then the cycles $c_1, \ldots, c_m$ have no symbols in common — they are *disjoint* cycles; and the product $c_1 c_2 \cdots c_m$ has the same effect as f on every element of $\{1, \ldots, n\}$, so $f = c_1 c_2 \cdots c_m$. ∎

The expression for a permutation f as a product of disjoint cycles is called the *cycle notation* for f. This expression is not quite unique. First, each cycle can begin with any one of its symbols — for example, the cycle (13275) has exactly the same effect on each of its symbols as the cycle (32751) or the cycle (75132). Second, the order in which we write the disjoint cycles does not matter — for example, $(124)(35)$ is the same permutation as $(35)(124)$. [But beware: it is only for *disjoint* cycles that the order does not matter; if you write down two non-disjoint cycles, their product one way round will be different from their product the other way round — e.g., $(123)(24) \neq (24)(123)$.]

Apart from these two ways of changing the cycles, the cycle notation for a permutation is unique. So, for example, the only ways of expressing the permutation $(124)(35) \in S_5$ as a product of disjoint cycles are

$$(124)(35) = (241)(35) = (412)(35) = (124)(53) =$$
$$(241)(53) = (412)(53) = (35)(124) = (35)(241) =$$
$$(35)(412) = (53)(124) = (53)(241) = (53)(412).$$

Multiplication of permutations is quite easy to do in your head, using the cycle notation. Here is an example.

Example 20.4

Let $f = (1325)(46) \in S_6$ and $g = (24)(163)(5) \in S_6$. Then

$$fg = (145)(26)(3).$$

I did this in my little head — honest! Here's what I said to myself: "Well, start with 1; g sends 1 to 6 and f sends 6 to 4, so fg sends 1 to 4; then it sends 4 to 5; then 5 to 1. So the first cycle of fg is (145). Now look at 2, which is not in this cycle: fg sends 2 to 6, then 6 to 2; so

(26) is the next cycle. Finally, fg sends 3 to 3, so (3) is the last cycle."
(Better stop talking to myself; people are looking at me strangely.)

If $g \in S_n$ is a permutation given in cycle notation, the *cycle-shape* of g is the sequence of numbers we get by writing down the lengths of the disjoint cycles of g in decreasing order. For example, the cycle-shape of the permutation $(1\,6\,3)(2\,4)(5\,8)(7)(9)$ in S_9 is $(3,2,2,1,1)$; normally we collect the repeated numbers and write this more succinctly as $(3,2^2,1^2)$.

Example 20.5
(1) How many permutations of cycle-shape $(2,1^{n-2})$ are there in S_n?

Answer This is easy: it is just the number of choices of a pair $\{i,j\}$ to make a 2-cycle $(i\,j)$, which is $\binom{n}{2}$.

(2) How many permutations of cycle-shape $(3,2^2,1)$ are there in S_8?

Answer This is slightly more tricky. First we have to choose three symbols i,j,k to put in a 3-cycle; and given i,j,k we can make two distinct 3-cycles — $(i\,j\,k)$ and $(i\,k\,j)$. So there are $\binom{8}{3} \times 2$ choices for the 3-cycle. Next, we choose two symbols l,m from the remaining five to put in the first 2-cycle, then two more, n,o to put in the second. This gives $\binom{5}{2} \times \binom{3}{2}$ choices, but we must divide this number by 2 since $(l\,m)(n\,o)$ is the same as $(n\,o)(l\,m)$. Hence the number of permutations of cycle-shape $(3,2^2,1)$ is

$$\binom{8}{3} \times 2 \times \binom{5}{2} \times \binom{3}{2} \times \frac{1}{2} = 1680.$$

Repeating a Permutation

Suppose f is the cycle $(1\,2\,3\,4\,5) \in S_5$. If we do f five times, we send 1 all the way round and back to 1, similarly 2 to 2, and so on. So f^5 sends $1 \to 1, 2 \to 2$, $\dots, 5 \to 5$; in other words, $f^5 = \iota$, the identity permutation.

In general, we define the *order* of a permutation $g \in S_n$ to be the smallest positive integer r such that $g^r = \iota$. In other words, the order of g is the smallest number of times we have to do g to send everything back to where it came from.

Orders of permutations are easy to work out using the cycle notation. We see from a few lines above that the order of a 5-cycle is equal to 5, and similarly that for any positive integer r, the order of an r-cycle is r (where an r-cycle just means a cycle of length r). Now consider a permutation with more than one cycle in its cycle notation, for example,

$$g = (243)(1658)(7) \in S_8.$$

To raise g to a power r and get $g^r = \iota$, we have to make r a multiple of 3 [to "kill off" the 3-cycle (243)] and also a multiple of 4 [to "kill off" the 4-cycle (1658)]. So r must be at least 12. Since it doesn't matter in which order we write the disjoint cycles of g, we see that

$$g^{12} = (243)(1658)(7) \cdot (243)(1658)(7) \cdot (243)(1658)(7) \cdots\cdots$$

$$= (243)^{12}(1658)^{12}(7)^{12} = \iota.$$

It follows that the order of g is equal to 12. In general, this reasoning shows the following.

PROPOSITION 20.4
The order of a permutation in cycle notation is equal to the least common multiple of the lengths of the cycles.

Example 20.6
How many permutations of order 2 are there in S_5?

Answer By Proposition 20.4, the permutations of order 2 in S_5 are those of cycle-shape $(2, 1^3)$ or $(2^2, 1)$. Arguing as in Example 20.5, we see that the number of permutations of cycle-shape $(2, 1^3)$ is $\binom{5}{2} = 10$, while the number of cycle-shape $(2^2, 1)$ is $\binom{5}{2} \times \binom{3}{2} \times \frac{1}{2} = 15$. So the total number of permutations of order 2 in S_5 is 25.

There are many situations in which it is useful to be able to work out the order of a permutation. Here is one example.

Example 20.7
A pack of eight cards is shuffled in the following way: the pack is divided into two equal parts and then "interlaced," so that if the original order was $1, 2, 3, 4, \ldots$, the new order is $1, 5, 2, 6, \ldots$. How many times must this shuffle be repeated before the cards are again in the original order?

Answer The shuffle gives the following permutation of the eight card positions:

$$f = \begin{pmatrix} 1 & 2 & 3 & 4 & 5 & 6 & 7 & 8 \\ 1 & 5 & 2 & 6 & 3 & 7 & 4 & 8 \end{pmatrix}.$$

In cycle notation, $f = (1)(253)(467)(8)$. This permutation has order 3, so the cards return to their original order after 3 shuffles.

Things get quite interesting if we consider the same problem for different numbers of cards — see Exercise 4 at the end of the chapter.

Even and Odd Permutations

We conclude this chapter by introducing an aspect of permutations that is a little more subtle than what we have seen so far.

Example 20.8

Let me begin with an example. Take $n = 3$ and let x_1, x_2, x_3 be three variables. We'll let the permutations in S_3 move around these variables just as they move around the numbers $1, 2, 3$. So, for instance, the permutation (132) sends $x_1 \to x_3$, $x_2 \to x_1$, $x_3 \to x_2$. Now define the expression

$$\Delta = (x_1 - x_2)(x_1 - x_3)(x_2 - x_3).$$

We can apply permutations in S_3 to Δ in an obvious way: for example, (123) sends Δ to $(x_2 - x_3)(x_2 - x_1)(x_3 - x_1)$. Notice that this is just the expression for Δ with two of the brackets, $(x_1 - x_2)$ and $(x_1 - x_3)$, reversed. So (123) sends $\Delta \to \Delta$. However, if we apply $(12)(3)$ to Δ, we end up with $(x_2 - x_1)(x_2 - x_3)(x_1 - x_3) = -\Delta$.

You can see that each permutation in S_3 sends Δ to either $+\Delta$ or $-\Delta$. Here's a table recording which signs occur:

g	$g(\Delta)$
ι	$+\Delta$
$(12)(3)$	$-\Delta$
$(13)(2)$	$-\Delta$
$(23)(1)$	$-\Delta$
(123)	$+\Delta$
(132)	$+\Delta$

We shall call those permutations that send Δ to $+\Delta$ *even permutations* and those that send Δ to $-\Delta$ *odd permutations*. So $\iota, (123)$ and (132) are even, while $(12)(3), (13)(2), (23)(1)$ are odd.

The definition of even and odd permutations for general n is very similar to the $n = 3$ example. Let $x_1, \ldots, x_n$ be variables, and take permutations in S_n

to move these variables in just the same way they move the symbols $1,\dots,n$ around. Define Δ to be the product of all $x_i - x_j$ for $i < j$. The notation for this is

$$\Delta = \prod_{1 \le i < j \le n} (x_i - x_j).$$

[The symbol $\prod$ means the product of all the terms $(x_i - x_j)$.] Just as in the example, we can apply any permutation $g \in S_n$ to Δ and the result will be either $+\Delta$ or $-\Delta$. Define the *signature* of g to be the number $sgn(g) \in \{+1, -1\}$ such that

$$g(\Delta) = sgn(g)\,\Delta.$$

This defines the signature function $sgn : S_n \to \{+1, -1\}$.

DEFINITION   A permutation $g \in S_n$ is an *even permutation* if $sgn(g) = +1$, and is an *odd permutation* if $sgn(g) = -1$.

It is not immediately obvious how to quickly calculate the signature of an arbitrary permutation. Certainly it would be a bit of a pain if we had to work out what $g(\Delta)$ was every time we wanted to know $sgn(g)$. Fortunately there is a somewhat more clever method available. It is based on the following result. In the statement of part (ii), a 2-cycle (ab) just means the permutation in S_n that swaps a and b and sends everything else to itself.

PROPOSITION 20.5
(i) The signature of the identity permutation, $sgn(\iota) = +1$.
(ii) For any $g, h \in S_n$, we have $sgn(gh) = sgn(g)\,sgn(h)$.
(iii) For any $g \in S_n$, we have $sgn(g^{-1}) = sgn(g)$.
(iv) The signature of any 2-cycle (ab) is -1.

PROOF   (i) Clearly $\iota(\Delta) = \Delta$, so $sgn(\iota) = +1$.

(ii) By definition of the signature function, we know that $gh(\Delta) = sgn(gh)\,\Delta$. But we also know that

$$gh(\Delta) = g(h(\Delta)) = g(sgn(h)\,\Delta) = sgn(g)\,sgn(h)\,\Delta.$$

So $sgn(gh) = sgn(g)\,sgn(h)$.

(iii) Using (i) and (ii) we see that $1 = sgn(\iota) = sgn(gg^{-1}) = sgn(g)sgn(g^{-1})$. Hence $sgn(g^{-1}) = sgn(g)$.

(iv) Let $t = (ab)$. We may as well assume that $a < b$. Then the brackets $(x_i - x_j)$ in Δ that appear reversed in $t(\Delta)$ (i.e., when x_a and x_b are swapped) are

$$(x_a - x_{a+1}), (x_a - x_{a+2}), \dots, (x_a - x_b),$$

$$(x_{a+1} - x_b), (x_{a+2} - x_b), \ldots, (x_{b-1} - x_b).$$

There are $b - a$ brackets in the first row and $b - a - 1$ in the second, making a total of $2b - 2a - 1$ reversed brackets. Since this is an odd number, $t(\Delta)$ must be $-\Delta$, so $sgn(t) = -1$. ■

We can use this result to work out the signature of any cycle.

PROPOSITION 20.6
The signature of any r-cycle in S_n is equal to $(-1)^{r-1}$.

PROOF Observe that a typical r-cycle $(a_1 a_2 \ldots a_r)$ can be written as a product of 2-cycles in the following way:

$$(a_1 a_2 \ldots a_r) = (a_1 a_r)(a_1 a_{r-1}) \cdots (a_1 a_2).$$

To see this, just note that the product on the right-hand side sends $a_1 \to a_2$, $a_2 \to a_3$, $\ldots a_r \to a_1$ and fixes everything else, which is exactly what the r-cycle $(a_1 a_2 \ldots a_r)$ on the left-hand side does. So the two sides are equal.

Now use Proposition 20.5: by part (ii), the signature of the r-cycle $(a_1 a_2 \ldots a_r)$ is equal to the product of the signatures of the 2-cycles on the right-hand side of the above equation, of which there are $r - 1$. By part (iv) of the proposition, each of these has signature -1. Hence the r-cycle has signature $(-1)^{r-1}$. ■

Using Proposition 20.6, it is easy to find the signature of any permutation $g \in S_n$. Let the cycle-shape of g be $(r_1, r_2, \ldots, r_k)$, so g is a product of disjoint cycles of lengths $r_1, \ldots, r_k$. Then by Proposition 20.5(ii), $sgn(g)$ is the product of the signatures of these cycles, so by Proposition 20.6, we have the following.

PROPOSITION 20.7
If $g \in S_n$ has cycle-shape $(r_1, r_2, \ldots, r_k)$, then

$$sgn(g) = (-1)^{r_1-1}(-1)^{r_2-1} \cdots (-1)^{r_k-1}.$$

So g is an even permutation if and only if the number of r_i that are even is an even number.

Example 20.9
(1) The permutation $(1234)(567)(89)(10\,11) \in S_{11}$ is odd.

(2) The permutations in S_5 that are even are precisely those that have cycle-shape (1^5), $(2^2,1)$, $(3,1^2)$ or (5). Using the counting methods illustrated in Example 20.5, we see that the total number of permutations in S_5 of each of these cycle-shapes is

cycle-shape	number
(1^5)	1
$(2^2,1)$	15
$(3,1^2)$	20
(5)	24
total	60

Notice that the total number of even permutations is 60, which is equal to $\frac{1}{2}(5!) = \frac{1}{2}|S_5|$. In Exercise 5 at the end of the chapter you are asked to prove that this is a general phenomenon — for any n, exactly half of the $n!$ permutations in S_n are odd and half are even.

You might ask what the point of this complicated definition of even and odd permutations is. For the most part, the answer is that you will see these cropping up in several more advanced topics in algebra later in your studies. For example, in group theory, the $\frac{1}{2}(n!)$ even permutations form a very important "subgroup" of S_n known as the *alternating group* (see Chapter 26). Another topic in which even and odd permutations play a key role is the theory of determinants of $n \times n$ matrices.

For now, let me merely offer the following amusing example to show a use for even and odd permutations.

Example 20.10
The "Fifteen Puzzle"

This puzzle consists of 15 square blocks labelled $1, 2, \ldots, 15$ arranged in a 4×4 frame, with one space, like this:

1	2	3	4
5	6	7	8
9	10	11	12
13	14	15	

To make a move you slide one block into the space, thereby creating a new space.

The problem is this: can you make a sequence of moves to change the above configuration to this one:

15	14	13	12
11	10	9	8
7	6	5	4
3	2	1	

Answer The answer is no, you can't. Here's how to prove this. Denote the space by □ so that the initial arrangement is

$$1\,2\,3\,4\,5\,6\,7\,8\,9\,10\,11\,12\,13\,14\,15\,\square.$$

Moving a number x into the space corresponds to doing the 2-cycle $(x\,\square)$. If we do a sequence of moves and end up with an arrangement with □ in its original place, then □ must have been moved upwards the same number of times as downwards, and leftwards the same number of times as rightwards. So the total number of moves must be even. Since each move is a 2-cycle, this means that the effect of the sequence of moves is a permutation that is a product of an even number of 2-cycles. By Proposition 20.5, this must therefore be an *even* permutation. However, we are looking for a sequence of moves that effects the permutation

$$\begin{pmatrix} 1 & 2 & 3 & 4 & 5 & 6 & 7 & 8 & 9 & 10 & 11 & 12 & 13 & 14 & 15 \\ 15 & 14 & 13 & 12 & 11 & 10 & 9 & 8 & 7 & 6 & 5 & 4 & 3 & 2 & 1 \end{pmatrix}.$$

In cycle notation this is $(1\,15)\,(2\,14)\,(3\,13)\,(4\,12)\,(5\,11)\,(6\,10)\,(7\,9)\,(8)$. But this is a product of seven 2-cycles, and hence is an odd permutation. Therefore, there is no sequence of moves that can effect this permutation.

The argument above showed that if a permutation of the blocks $1, 2, \ldots, 15$ can be achieved by a sequence of moves, then it must be an even permutation. The question of whether *every* even permutation can be achieved is much more subtle. If you are interested in reading further about this, have a look at the article by A. Archer, *A modern treatment of the 15 puzzle*, American Math. Monthly, Vol.106 (1999), pp.793–799.

Exercises for Chapter 20

1. Complete the multiplication table for S_3 started in Example 20.3.

2. Let f and g be the following permutations in S_7:

$$f = \begin{pmatrix} 1 & 2 & 3 & 4 & 5 & 6 & 7 \\ 3 & 1 & 5 & 7 & 2 & 6 & 4 \end{pmatrix}, \quad g = \begin{pmatrix} 1 & 2 & 3 & 4 & 5 & 6 & 7 \\ 3 & 1 & 7 & 6 & 4 & 5 & 2 \end{pmatrix}.$$

Write down in cycle notation the permutations $f, g, g^2, g^3, f \circ g, (f \circ g)^{-1}$ and $g^{-1} \circ f^{-1}$.

What is the order of f? What is the order of $f \circ g$?

3. (a) List the numbers that occur as the orders of elements of S_4, and calculate how many elements there are in S_4 of each of these orders.

   (b) List all the possible cycle-shapes of even permutations in S_6.

   (c) Calculate the largest possible order of any permutation in S_{10}.

   (d) Calculate the largest possible order of any even permutation in S_{10}.

   (e) Find a value of n such that S_n has an element of order greater than n^2.

4. A pack of $2n$ cards is shuffled by the "interlacing" method described in Example 20.7 — in other words, if the original order is $1, 2, 3, \ldots, 2n$, the new order after the shuffle is $1, n+1, 2, n+2, \ldots, n, 2n$. Work out how many times this shuffle must be repeated before the cards are again in the original order in the following cases:

   (a) $n = 10$

   (b) $n = 12$

   (c) $n = 14$

   (d) $n = 16$

   (e) $n = 24$

   (f) $n = 26$ (i.e., a real pack of cards).

   Investigate this question as far as you can for general n — it is quite fascinating!

5. Prove that exactly half of the $n!$ permutations in S_n are even.

   (*Hint:* Show that if g is an even permutation, then $g\,(12)$ is odd. Try to use this to define a bijection from the set of odd permutations to the set of even permutations.)

6. This question is about the 3×3 version of the Fifteen Puzzle of Example 20.10. Starting with the configuration

$$1\ 2\ 3$$
$$4\ 5\ 6$$
$$7\ 8\ \square$$

which of the following configurations can be reached by a sequence of moves?

$$
\begin{array}{ccc}
3\ 2\ 1 & 1\ \square\ 2 & 1\ 7\ 2 \\
4\ 5\ 6\ , & 3\ 4\ 5\ , & 6\ 4\ 5 \\
7\ 8\ \square & 6\ 7\ 8 & 3\ 8\ \square
\end{array}
$$

7. Let S be a set of size m and T a set of size n. Assume that $m \geq n$. This question is about the number of onto functions from S to T, which is much more complicated than the corresponding question about 1-1 functions (see Exercise 7 of Chapter 19).

(a) What is the number of onto functions from S to T if $m = n$?

(b) Show that if $m = n + 1$, the number of onto functions from S to T is

$$\binom{n+1}{2} \cdot n!$$

(c) Show that if $m = n + 2$, the number of onto functions from S to T is

$$\binom{n+2}{3} \cdot n! + \binom{n+2}{n-2,2,2} \cdot n!$$

(*Hint:* An onto function $S \to T$ will either send some set of 3 elements of S to the same element of T, or send two pairs of elements of S to two elements of T. Count the numbers of such functions separately in these two cases.)

8. Critic Ivor Smallbrain has been engaged for the prestigious role of dressing up as Father Christmas at Harrods this year. There, he will have to distribute $n + 3$ toys to n children. He must make sure that every child gets at least one toy, but he can give the extra three toys to any of the children.

How many ways are there in which Ivor can distribute the toys?

(*Note:* This is just the number of onto functions from a set of size $n + 3$ to a set of size n, if that's any help.)

Chapter 21

Infinity

Given two finite sets, it is simple to compare their sizes. For example, we would say that the set of corners of a pentagon is larger than the set of players in a string quartet, simply because the first set has five elements, while the second has only four.

But can we compare the sizes of *infinite* sets in any meaningful way? We have encountered many different infinite sets at various points in this book, such as $\mathbb{N}, \mathbb{Z}, \mathbb{Q}, \mathbb{R}, \mathbb{C}, \mathbb{N} \times \mathbb{N}, \mathbb{Q} \times \mathbb{R} \times \mathbb{C}$, and so on. How can we compare these with each other?

There is a way to do this using functions. To set this up, let us begin with an elementary observation about finite sets. If S is a set of size n, say $S = \{s_1, s_2, \ldots, s_n\}$, then the function $f : S \to \{1, 2, \ldots, n\}$ defined by

$$f(s_1) = 1, f(s_2) = 2, \ldots, f(s_n) = n.$$

is a bijection. Thus we can say

$$S \text{ has size } n \iff \text{ there is a bijection from } S \text{ to } \{1, 2, \ldots, n\}.$$

We now extend this notion to arbitrary sets.

DEFINITION *Two sets A and B are said to be* equivalent *to each other if there is a bijection from A to B. We write $A \sim B$ if A and B are equivalent to each other.*

In accordance with the preamble to the definition, we can informally think of two sets that are equivalent to each other as "having the same size."

Before doing anything else, let us establish that the relation $\sim$ is an equivalence relation on sets.

PROPOSITION 21.1
The relation $\sim$ defined above is an equivalence relation.

PROOF   First we show $\sim$ is reflexive; that is, $A \sim A$ for any set A. This is true since the identity function, $\iota_A : A \to A$ defined by $i_A(a) = a$ for all $a \in A$, is a bijection.

Next we show $\sim$ is symmetric. Suppose $A \sim B$, so there is a bijection $f : A \to B$. Then the inverse function $f^{-1} : B \to A$ is a bijection, so $B \sim A$.

Finally, we show $\sim$ is transitive. Suppose $A \sim B$ and $B \sim C$, so there are bijections $f : A \to B$ and $g : B \to C$. Then, by Proposition 19.2, $g \circ f : A \to C$ is a bijection, so $A \sim C$. Hence $\sim$ is transitive.   ∎

Example 21.1

(1) Let $A = \mathbb{N}$ and let $B = \{2n \,|\, n \in \mathbb{N}\}$, the set of all positive even numbers. Then the function $f : A \to B$ defined by

$$f(n) = 2n \quad \text{for all } n \in \mathbb{N}$$

is a bijection. Thus $A \sim B$; i.e., $\mathbb{N} \sim$ even numbers in $\mathbb{N}$.

This example shows that $\mathbb{N}$ can be equivalent to a subset of itself. (Informally, $\mathbb{N}$ "has the same size" as a subset of itself.) It is of course not possible for any *finite* set to have this property.

(2) Suppose A is a set that is equivalent to $\mathbb{N}$. This means there is a bijection $f : \mathbb{N} \to A$. For $n \in \mathbb{N}$, let $f(n) = a_n \in A$. Since f is onto, we then have

$$A = \{a_1, a_2, a_3, \ldots, a_n, \ldots\}.$$

In other words, we can *list* all the elements of A as $a_1, a_2, a_3, \ldots$.

Countable Sets

The listing property of the last example is so fundamental that we give it a special definition.

DEFINITION   *A set A is said to be* countable *if A is equivalent to $\mathbb{N}$. In other words, A is countable if it is an infinite set, all of whose elements can be listed as $A = \{a_1, a_2, a_3, \ldots, a_n, \ldots\}$.*

Example 21.2

(1) $\mathbb{N}$ is obviously itself countable: its elements can be listed as $1, 2, 3, \ldots$.

(2) The set $B = \{2n \,|\, n \in \mathbb{N}\}$ of positive even numbers is countable: the elements of B can be listed a $2, 4, 6, 8, \ldots$.

(3) What about $\mathbb{Z}$ — is it countable? This is not quite so obvious, but the answer is yes because we can list all the elements of $\mathbb{Z}$ as $0, 1, -1, 2, -2, 3, -3, \ldots$. Correspondingly, we could define a bijection $f : \mathbb{N} \to \mathbb{Z}$ by

$$f(2n) = n \quad \text{and} \quad f(2n-1) = -(n-1)$$

for all $n \geq 1$.

The same idea shows that the union of any two countable sets is countable (just list the elements alternately, omitting any repetitions).

The next proposition provides us with many more examples of countable sets.

PROPOSITION 21.2
Every infinite subset of $\mathbb{N}$ is countable.

PROOF   Let S be an infinite subset of $\mathbb{N}$. Take s_1 to be the smallest integer in S; then take s_2 to be the smallest integer in $S - \{s_1\}$, take s_3 to be the smallest in $S - \{s_1, s_2\}$, and so on. In this way, we list all the elements of S in ascending order as $S = \{s_1, s_2, s_3, \ldots\}$. Therefore, S is countable. (A bijection $f : \mathbb{N} \to S$ would simply be $f(1) = s_1$, $f(2) = s_2$, $f(3) = s_3$, and so on.)   ∎

Let us now consider the question of whether $\mathbb{Q}$, the set of rationals, is countable. This is much more subtle than any of the previous examples. For a start, we certainly cannot list the positive rationals in ascending order, since, no matter what rational x we started the list with, there would be a smaller one (e.g., $\frac{1}{2}x$) that would then not appear on the list. However, could it be possible to devise a devilishly clever alternative way to list the rationals?

Somewhat amazingly, the answer is yes.

PROPOSITION 21.3
The set of rationals $\mathbb{Q}$ is countable.

PROOF   First consider $\mathbb{Q}^+$, the set of positive rationals. We show how to list the elements of $\mathbb{Q}^+$. The key is first to write the positive rationals in an array as follows:

$$\begin{array}{ccccc}
\dfrac{1}{1} & \dfrac{2}{1} & \dfrac{3}{1} & \dfrac{4}{1} & \dfrac{5}{1} \quad \cdots \\[2mm]
\dfrac{1}{2} & \dfrac{2}{2} & \dfrac{3}{2} & \dfrac{4}{2} & \dfrac{5}{2} \quad \cdots \\[2mm]
\dfrac{1}{3} & \dfrac{2}{3} & \dfrac{3}{3} & \dfrac{4}{3} & \dfrac{5}{3} \quad \cdots \\[2mm]
\dfrac{1}{4} & \dfrac{2}{4} & \dfrac{3}{4} & \dfrac{4}{4} & \dfrac{5}{4} \quad \cdots \\[2mm]
\dfrac{1}{5} & \dfrac{2}{5} & \dfrac{3}{5} & \dfrac{4}{5} & \dfrac{5}{5} \quad \cdots \\
\cdot & \cdot & \cdot & \cdot & \cdot \quad \cdots \\
\cdot & \cdot & \cdot & \cdot & \cdot \quad \cdots
\end{array}$$

Draw a zig-zag line through this array as follows:

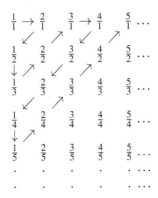

We can now list the positive rationals by simply moving along the zig-zag line in the direction of the arrows, writing down each number as we reach it (and omitting numbers we have already written down, such as $\frac{2}{2}, \frac{3}{3}, \frac{4}{2}$ and so on). The list starts like this:

$$1, 2, \frac{1}{2}, \frac{1}{3}, 3, 4, \frac{3}{2}, \frac{2}{3}, \frac{1}{4}, \frac{1}{5}, 5, \cdots.$$

Thus we obtain a complete list of all the positive rationals, showing that $\mathbb{Q}^+$ is countable.

Finally, we need to deduce that $\mathbb{Q}$ is countable. Let the above list of the elements of $\mathbb{Q}^+$ be $\mathbb{Q}^+ = \{q_1, q_2, q_3, \ldots\}$. Then we can list the elements of $\mathbb{Q}$ as

$$\mathbb{Q} = \{0, q_1, -q_1, q_2, -q_2, q_3, -q_3, \ldots\},$$

which shows that $\mathbb{Q}$ is countable. ∎

The next proposition provides a quite useful method for showing that sets are countable.

PROPOSITION 21.4

Let S be an infinite set. If there is a 1-1 function $f : S \to \mathbb{N}$, then S is countable.

PROOF   Recall that the image of f is the set

$$f(S) = \{f(s) \mid s \in S\} \subseteq \mathbb{N}.$$

Since f is 1-1, $f(S)$ is an infinite set. Therefore, by Proposition 21.2, $f(S)$ is countable. Consequently, there is a bijection $g : \mathbb{N} \to f(S)$.

Now we can regard f as a function from S to $f(S)$. As such, f is onto; hence, as f is also 1-1, f is a bijection from S to $f(S)$. There is therefore an inverse function $f^{-1} : f(S) \to S$.

Finally, consider the composition $f^{-1} \circ g : \mathbb{N} \to S$. By Proposition 19.2(c), this is a bijection. This means that S is countable.   ∎

This proposition can be used in many further examples:

Example 21.3

(1) Here is another proof that $\mathbb{Q}^+$ is countable. Define $f : \mathbb{Q}^+ \to \mathbb{N}$ by

$$f\left(\frac{m}{n}\right) = 2^m 3^n$$

where $m, n \in \mathbb{N}$ and $\frac{m}{n}$ is in lowest terms. Then f is 1-1, since

$$f\left(\frac{m}{n}\right) = f\left(\frac{p}{q}\right) \;\Rightarrow\; 2^m 3^n = 2^p 3^q \;\Rightarrow\; m = p, n = q$$

using the Fundamental Theorem of Arithmetic 11.1. Hence $\mathbb{Q}^+$ is countable by Proposition 21.4.

(2) A very similar proof shows that the Cartesian product $\mathbb{N} \times \mathbb{N}$ is countable: just define $f : \mathbb{N} \times \mathbb{N} \to \mathbb{N}$ by $f(m, n) = 2^m 3^n$ and observe again that f is 1-1. Likewise, $\mathbb{N} \times \mathbb{N} \times \mathbb{N}$ is countable, since the function $g(m, n, l) = 2^m 3^n 5^l$ from $\mathbb{N} \times \mathbb{N} \times \mathbb{N} \to \mathbb{N}$ is 1-1; and so on — the Cartesian product of any finite number of copies of $\mathbb{N}$ is countable.

An Uncountable Set

We have shown that many sets are countable. Are there in fact any infinite sets that are *not* countable? The answer is yes. Here is the most famous example of an uncountable set — where, not surprisingly, an *uncountable* set is defined to be an infinite set that is not countable.

THEOREM 21.1
The set $\mathbb{R}$ of all real numbers is uncountable.

PROOF   We prove the theorem by contradiction. So suppose that $\mathbb{R}$ is countable. This means that we can list all the elements of $\mathbb{R}$ as

$$\mathbb{R} = \{r_1, r_2, r_3, \ldots\}.$$

Express each of the r_is in the list as a decimal:

$$r_1 = m_1.a_{11}a_{12}a_{13}\ldots$$
$$r_2 = m_2.a_{21}a_{22}a_{23}\ldots$$
$$r_3 = m_3.a_{31}a_{32}a_{33}\ldots$$
$$\cdot \quad \cdot$$
$$\cdot \quad \cdot$$
$$r_n = m_n.a_{n1}a_{n2}a_{n3}\ldots$$
$$\cdot \quad \cdot$$
$$\cdot \quad \cdot$$

(where each $m_i \in \mathbb{Z}$ and each $a_{ij} \in \{0,1,2,\ldots,9\}$).

Now define a real number $r = 0.b_1b_2b_3\ldots$ as follows.

To choose the first decimal digit b_1: if $a_{11} \neq 1$, let $b_1 = 1$; and if $a_{11} = 1$, let $b_1 = 2$. (Hence $b_1 \neq a_{11}$.)

To choose the second decimal digit b_2: if $a_{22} \neq 1$, let $b_2 = 1$; and if $a_{22} = 1$, let $b_2 = 2$. (Hence $b_2 \neq a_{22}$.)

And so on: in general, to choose the nth decimal digit b_n: if $a_{nn} \neq 1$, let $b_n = 1$; and if $a_{nn} = 1$, let $b_n = 2$. (Hence $b_n \neq a_{nn}$.)

In this way we define a real number $r = 0.b_1b_2b_3\ldots$. Since $r_1, r_2, r_3, \ldots$ is a list of *all* real numbers, r must belong to this list, so $r = r_n$ for some n. But $b_n \neq a_{nn}$, so r and r_n differ in their n^{th} decimal digit. Note also that r does not end in recurring 9s or 0s (all the b_is are 1 or 2). Hence $r \neq r_n$, which is a contradiction.

Therefore, there is no bijection from $\mathbb{N}$ to $\mathbb{R}$, which is to say that $\mathbb{R}$ is uncountable.   ∎

This famous theorem is due to Georg Cantor (1874), the founder of modern set theory. The wonderfully clever idea of the proof — defining the decimal $r = 0.b_1b_2b_3 \ldots$ by adjusting the "diagonally placed" decimal digits a_{nn} in the array of r_is — is called Cantor's "diagonal argument" and can be used to prove that all sorts of other sets are uncountable. (See Exercises 4 and 5 at the end of the chapter.)

A consequence of Theorem 21.1 is that the set of irrationals $\mathbb{R} - \mathbb{Q}$ is un-countable — for if it were countable, then $\mathbb{R}$ would be the union of two count-able sets and hence would be countable. Thus, there are in some sense "more" irrational numbers than there are rationals.

A Hierarchy of Infinities

DEFINITION   *Let A and B be sets. If A and B are equivalent to each other (i.e., there is a bijection from A to B), we say that A and B have the same* cardinality, *and write* $|A| = |B|$.

If there is a 1-1 function from A to B, we write $|A| \leq |B|$.

And if there is a 1-1 function from A to B, but no *bijection from A to B, we write* $|A| < |B|$, *and say that A has smaller cardinality than B. (Thus,* $|A| < |B|$ *is the same as saying that* $|A| \leq |B|$ *and* $|A| \neq |B|$.)

According to this definition, we have

$$|\mathbb{N}| = |\mathbb{Q}| = |\mathbb{N} \times \mathbb{N}|$$

and

$$|\mathbb{N}| < |\mathbb{R}|.$$

Thus there are at least two different types of "infinity," namely $|\mathbb{N}|$ and $|\mathbb{R}|$.

Are there more types of infinity? For example, is there a set of greater cardinality than $\mathbb{R}$?

The answer is yes, and again this is due to Cantor. To understand this, we first need a definition.

DEFINITION   *If S is a set, let P(S) be the set consisting of all the subsets of S.*

For example, if $S = \{1,2\}$ then $P(S) = \{\{1,2\}, \{1\}, \{2\}, \emptyset\}$. In general, if S is a finite set of size n, then $|P(S)| = 2^n$ by Proposition 17.4.

Cantor's theory is based on the following result.

PROPOSITION 21.5
Let S be a set. Then there is no bijection from S to P(S). Consequently,
$|S| < |P(S)|$.

Using the proposition, we obtain a hierarchy of infinities, starting at $|\mathbb{N}|$:

$$|\mathbb{N}| < |P(\mathbb{N})| < |P(P(\mathbb{N}))| < |P(P(P(\mathbb{N})))| < \cdots .$$

Thus there are indeed many types of "infinity."

PROOF   Here is a proof of Proposition 21.5. It is a very subtle proof by contradiction. You may well have to go through it a few times before really understanding it.

Suppose there is a bijection $f : S \to P(S)$. Then *every* subset of S is equal to $f(s)$ for some $s \in S$.

For any $s \in S$, $f(s)$ is a subset of S, and it is certainly the case that either $s \in f(s)$ or $s \notin f(s)$. [For example, there exists s_1 such that $f(s_1) = S$, and then $s_1 \in f(s_1)$; likewise, there exists s_2 such that $f(s_2) = \emptyset$, and then $s_2 \notin f(s_2)$.] Define A to be the set of all elements s of S such that $s \notin f(s)$; symbolically,

$$A = \{s \in S \mid s \notin f(s)\}.$$

(In the above notation, $s_1 \notin A$ but $s_2 \in A$.)

Certainly A is a subset of S; that is, $A \in P(S)$. Therefore, as f is a bijection, $A = f(a)$ for some $a \in S$.

We now ask the question: does a belong to A?

If $a \notin A$, then $a \notin f(a)$, so by definition of A, we have $a \in A$. This is a contradiction. And if $a \in A$ then $a \in f(a)$, so by definition of A we have $a \notin A$, again a contradiction.

Thus we have reached a contradiction in any case. So we conclude that there cannot be a bijection from S to $P(S)$.

Since there is certainly a 1-1 function $f : S \to P(S)$, namely $f(s) = \{s\}$ for all $s \in S$, it follows that $|S| < |P(S)|$, and the proof is complete. ∎

Exercises for Chapter 21

1. (a) Show that if A is a countable set and B is a finite set, then $A \cup B$ is countable.

(b) Show that if A and B are both countable sets, then $A \cup B$ is countable.

2. (a) Show that if each of the sets S_n $(n = 1, 2, 3, \ldots)$ is countable, then the union $S = \bigcup_{n=1}^{\infty} S_n$ is also countable.

   (b) Show that if S and T are countable sets, then the Cartesian product $S \times T$ is also countable. Hence show that $\bigcup_{n=1}^{\infty} S^n$ is countable, where $S^n = S \times S \times \cdots \times S$ (n times).

3. Write down a sequence $z_1, z_2, z_3, \ldots$ of complex numbers with the following property: for any complex number w and any positive real number ε, there exists N such that $|w - z_N| < \varepsilon$.

   (*Hint:* Try to use the fact that $\mathbb{Q}$ is countable.)

4. Let S be the set consisting of all infinite sequences of 0s and 1s (so a typical member of S is 010011011100110..., going on forever). Use Cantor's diagonal argument to prove that S is uncountable.

5. (a) Let S be the set consisting of all the finite subsets of $\mathbb{N}$. Prove that S is countable.

   (b) Let T be the set consisting of all the infinite subsets of $\mathbb{N}$. Prove that T is uncountable.

   (c) Prove that the set of all functions $f : \mathbb{N} \to \mathbb{N}$ is uncountable.

6. Every Tuesday, critic Ivor Smallbrain drinks a little too much, staggers out of the pub, and performs a kind of random walk towards his home. At each step of this walk, he stumbles either forwards or backwards, and the walk ends either when he collapses in a heap or when he reaches his front door (one of these always happens after a finite [possibly very large] number of steps). Ivor's Irish friend Gerry O'Laughing always accompanies him and records each random walk as a sequence of 0s and 1s: at each step he writes a 1 if the step is forwards and a 0 if it is backwards.

   Prove that the set of all possible random walks is countable.

Chapter 22

Introduction to Analysis: Bounds

In the next three chapters of the book — this one and the next two — I want to introduce you to a topic with a different flavour from the rest of the book. This is a topic called mathematical analysis, or just analysis for short. This is a huge area, which at an undergraduate level starts off with the study of the real and complex numbers, and functions defined on them. Of course we can't cover very much of the subject — that would require several more books — but we will do enough to prove several interesting results, such as putting the existence of decimal expressions for real numbers on a rigorous footing, and proving the existence of nth roots of positive real numbers (stated in Proposition 4.1).

Before we can start studying functions of the real numbers, we need to go more deeply into some basic properties of the real numbers themselves, and that's what we'll do in this chapter.

Upper and Lower Bounds

Our study will be based on the theory of *bounds* for sets of real numbers. Here is the definition.

DEFINITION   *Let S be a non-empty subset of $\mathbb{R}$. (So S is a set consisting of some real numbers, and $S \neq \emptyset$). We say that a real number u is an* upper bound *for S if*

$$s \leq u \ \text{for all} \ s \in S.$$

Likewise, l is a lower bound *for S if*

$$s \geq l \ \text{for all} \ s \in S.$$

Example 22.1
(1) Let

$$S = \left\{ \frac{1}{n} \,\middle|\, n \in \mathbb{N} \right\} = \left\{ 1, \frac{1}{2}, \frac{1}{3}, \frac{1}{4}, \dots \right\}.$$

Then 1 is an upper bound for S; so are 2, 17 and indeed any number that is at least 1. Also 0 is a lower bound for S, and so is any number less than or equal to 0.

(2) If $S = \mathbb{Z}$, then S has no upper or lower bound.

(3) If

$$S = \left\{ x \,\middle|\, x \in \mathbb{Q}, x^2 < 2 \right\}$$

(i.e., the set of rationals with square less than 2), then $\sqrt{2}$ is an upper bound for S and $-\sqrt{2}$ is a lower bound.

As we see from these examples, a set can have many upper bounds. It turns out to be a fundamental question to ask whether, among all the upper bounds, there is always a least one. Let us first formally define such a thing.

DEFINITION   *Let S be a non-empty subset of $\mathbb{R}$, and suppose S has an upper bound. We say that a real number c is a least upper bound for S (abbreviated LUB), if the following two conditions hold:*

(i) c is an upper bound for S, and
(ii) if u is any other upper bound for S, then $u \geq c$.

Similarly, d is a greatest lower bound (GLB) for S if

(a) d is a lower bound for S, and
(b) if l is any other lower bound for S, then $l \leq d$.

Example 22.2
Let $S = \{ \frac{1}{n} \mid n \in \mathbb{N} \}$ as in Example 22.1(1).

We claim that 1 is a LUB for S. To see this, observe that 1 is an upper bound; and any other upper bound is at least 1, since $1 \in S$.

We also claim that 0 is a GLB for S. This is not quite so obvious. First, 0 is a lower bound. Let l be another lower bound for S. If $l > 0$, then we can find $n \in \mathbb{N}$ such that $\frac{1}{n} < l$; but $\frac{1}{n} \in S$, so this is not possible as l is a lower bound for S. Hence $l \leq 0$, which proves that 0 is a GLB for S.

It is an absolutely fundamental property of the real numbers that every set which has an upper bound also has a least upper bound. We won't prove this here, as it relies on a rigorous construction of the real numbers, which we have not done. It is known as the Completeness Axiom for $\mathbb{R}$:

COMPLETENESS AXIOM   Let S be a non-empty subset of $\mathbb{R}$.
    (I) If S has a lower bound, then it has a greatest lower bound.
    (II) If S has an upper bound, then it has a least upper bound.

For interested readers, I have included Exercise 8 at the end of the chapter, which leads you through the rigorous construction of the real numbers from the rationals and proves the Completeness Axiom.

If a set S has a LUB, it is easy to see that it has only one LUB. We leave this to the reader (Exercise 3 at the end of the chapter). So it makes sense to talk about *the* least upper bound of S. We sometimes denote this by LUB(S). Likewise, a set S with a lower bound has only one GLB, denoted by GLB(S).

As we have said, the Completeness Axiom underlies the whole of the theory of the real numbers, and you will see it used many times in your future study of mathematics. In the next two chapters we'll apply it in the study of limits and functions and use it to justify the existence of decimal expressions and n^{th} roots. For now, here is a little example to give you the flavour of how the axiom can be used.

Example 22.3

We saw in Chapter 2 how to prove the existence of the real number $\sqrt{2}$, and more generally of $\sqrt{n}$ for any positive integer n, by means of a clever geometrical construction. However, proving the existence of the real cube root of 2 is not so easy. In this example I'll show you how to use the Completeness Axiom to prove the existence of $2^{1/3}$.

The key idea is to define the following set of real numbers:

$$S = \left\{ x \mid x \in \mathbb{R}, x^3 < 2 \right\}.$$

Thus, S is the set of all real numbers whose cube is less than 2.

First note that S has an upper bound; for example, 2 is an upper bound, since if $x^3 < 2$ then $x < 2$. Therefore, by the Completeness Axiom, S has a least upper bound. Say

$$c = \text{LUB}(S).$$

We shall show that $c^3 = 2$. We do this by contradiction.

Assume then that $c^3 \neq 2$. Then either $c^3 < 2$ or $c^3 > 2$. We consider these two possibilities separately, in each case obtaining a contradiction.

Case 1 Assume that $c^3 < 2$. Our strategy in this case is to find a small number $\alpha > 0$ such that $(c+\alpha)^3 < 2$ still; this will mean that $c+\alpha \in S$, whereas c is an upper bound for S, a contradiction.

To find α, we argue as follows. (I have chosen to present the steps "in reverse" in order to make it clear how the argument was found.) We have

$$(c+\alpha)^3 < 2 \Leftarrow c^3 + 3c^2\alpha + 3c\alpha^2 + \alpha^3 < 2$$
$$\Leftarrow 2 - c^3 > 3c^2\alpha + 3c\alpha^2 + \alpha^3$$
$$\Leftarrow 2 - c^3 > \alpha\left(3c^2 + 3c + 1\right) \text{ and } 0 < \alpha < 1.$$

(The last inequality follows, since when $0 < \alpha < 1$, we have $\alpha^2 < \alpha$ and $\alpha^3 < \alpha$.) Since $2 - c^3 > 0$, we can choose α such that

$$0 < \alpha < 1 \text{ and } \alpha < \frac{2 - c^3}{3c^2 + 3c + 1}.$$

Then, by the above implications it follows that $(c+\alpha)^3 < 2$. This leads to a contradiction, as explained before.

Case 2 Now assume that $c^3 > 2$. In this case our strategy is to find a small number $\beta > 0$ such that $(c-\beta)^3 > 2$ still. If we do this, then for $x \in S$ we have $x^3 < 2 < (c-\beta)^3$, hence $x < c - \beta$, and so $c - \beta$ is an upper bound for S. However, c is the LUB of S, so this is a contradiction.

To find β, note that

$$(c-\beta)^3 > 2 \Leftarrow c^3 - 2 > 3c^2\beta - 3c\beta^2 + \beta^3$$
$$\Leftarrow c^3 - 2 > \beta(3c^2 + 3c + 1) \text{ and } 0 < \beta < 1.$$

Since $c^3 - 2 > 0$ we can choose β such that

$$0 < \beta < 1 \text{ and } \beta < \frac{c^3 - 2}{3c^2 + 3c + 1}.$$

Then, by the above implications, it follows that $(c-\beta)^3 > 2$. This leads to a contradiction, as explained before.

Thus we have reached a contradiction in both Cases 1 and 2, from which we conclude that $c^3 = 2$. In other words, c is the real cube root of 2.

You can see that the calculations in this example are both tricky and tedious. In Chapter 24 we shall prove a general result (the Intermediate Value Theorem) that makes this kind of argument easy, smooth and free of tricky or tedious calculations.

Exercises for Chapter 22

1. Which of the following sets S have an upper bound and which have a lower bound? In the cases where these exist, state what the least upper bounds and greatest lower bounds are.

   (i) $S = \{-1, 3, 7, -2\}$.

   (ii) $S = \{x \mid x \in \mathbb{R} \text{ and } |x - 3| < |x + 7|\}$.

   (iii) $S = \{x \mid x \in \mathbb{R} \text{ and } x^3 - 3x < 0\}$.

   (iv) $S = \{x \mid x \in \mathbb{N} \text{ and } x^2 = a^2 + b^2 \text{ for some } a, b \in \mathbb{N}\}$.

2. Write down proofs of the following statements about sets A and B of real numbers:

   (a) If x is an upper bound for A, and $x \in A$, then x is a least upper bound for A.

   (b) If $A \subseteq B$, then a lower bound for B is also a lower bound for A.

   (c) If $A \subseteq B$ and a greatest lower bound of A is x, and a greatest lower bound of B is y, then $y \leq x$.

3. Prove that if S is a set of real numbers, then S cannot have two different least upper bounds or greatest lower bounds.

4. Find the LUB and GLB of the following sets:

   (i) $\{x \mid x = 2^{-p} + 3^{-q} \text{ for some } p, q \in \mathbb{N}\}$

   (ii) $\{x \in \mathbb{R} \mid 3x^2 - 4x < 1\}$

   (iii) the set of all real numbers between 0 and 1 whose decimal expression contains no nines

5. (a) Find a set of rationals having rational LUB.

   (b) Find a set of rationals having irrational LUB.

   (c) Find a set of irrationals having rational LUB.

6. Which of the following statements are true and which are false?

   (a) Every set of real numbers has a GLB.

   (b) For any real number r, there is a set of rationals having GLB equal to r.

   (c) Let $S \subseteq \mathbb{R}, T \subseteq \mathbb{R}$, and define $ST = \{st \mid s \in S, t \in T\}$, the set of all products of elements of S with elements of T. If c is the GLB of S, and d is the GLB of T, then cd is the GLB of ST.

(d) If S is a set of real numbers such that $\text{GLB}(S) \notin S$, then S must be an infinite set.

7. Prove that the cubic equation $x^3 - x - 1 = 0$ has a real root (i.e., prove that there exists a real number c such that $c^3 - c - 1 = 0$). (*Hint:* Try to find c as the LUB of a suitable set.)

8. Here is an exercise, not for the faint-hearted, leading you through the rigorous construction of the real numbers from the rationals $\mathbb{Q}$ and proving the Completeness Axiom.

Call a subset S of $\mathbb{Q}$ a *Dedekind cut* if $S \neq \emptyset, \mathbb{Q}$ and S satisfies the following two conditions:

(i) for any $s \in S$, S contains all the rationals less than s

(ii) S has no maximum (i.e., there is no element of S which is greater than all other members of S).

(For example, $S = \{x \in \mathbb{Q} : x < \frac{1}{2}\}$ is a Dedekind cut, but $\{x \in \mathbb{Q} : x \leq \frac{1}{2}\}$ is not.)

Define $\mathbb{R}$ to be the set of all Dedekind cuts. We need to define addition, multiplication and ordering on $\mathbb{R}$ and show that it has all the basic properties such as Rules 2.1 and the Completeness Axiom.

Strangely, perhaps, it's easier to get the Completeness Axiom than the other things, so let's do this first. Define an ordering on $\mathbb{R}$ by simply saying that if S, T are Dedekind cuts, then $S < T$ if and only if $S \subset T$. Naturally we say that $S \leq T$ if either $S = T$ or $S < T$. With this definition, prove that the Completeness Axiom for $\mathbb{R}$ holds, as follows: let A be a subset of $\mathbb{R}$ with an upper bound U (so $S \leq U$ for all $S \in A$). Define $L = \bigcup_{S \in A} S$. Prove that $L \in \mathbb{R}$ (i.e., L is a Dedekind cut) and L is a least upper bound for A.

Now for addition, multiplication and so on. First identify $\mathbb{Q}$ with a subset of $\mathbb{R}$ by taking $q \in \mathbb{Q}$ to the Dedekind cut $S_q = \{x \in \mathbb{Q} : x < q\}$.

For two Dedekind cuts S, T, define their sum to be $S + T = \{s + t : s \in S, t \in T\}$. Show that this addition agrees with the usual addition on $\mathbb{Q} \subset \mathbb{R}$.

Similarly define multiplication on $\mathbb{R}$ and division on $\mathbb{R} - \{0\}$ and show that they agree with their usual definitions on $\mathbb{Q} \subset \mathbb{R}$. [This needs to be done rather carefully — for example, the obvious definition of multiplication $ST = \{st : s \in S, t \in T\}$ does not work at all. Here's a hint to get you started: for $S, T > 0$ (where 0 is identified with S_0 as above), define

$$ST = \mathbb{Q}_{\leq 0} \cup \{st : s \in S, t \in T \text{ with } s > 0, t > 0\}.]$$

Show finally that these definitions of $+, \times$ satisfy the Rules 2.1.

(*Hint:* You might find it helpful to note that if we already knew what the real numbers were, then the Dedekind cuts would just be $S_r = \{x \in \mathbb{Q} : x < r\}$ for $r \in \mathbb{R}$.)

9. Let $x_1, x_2, x_3, \ldots$ be a sequence of real numbers (going on forever). For any integer $n \geq 1$, define T_n to be the set $\{x_n, x_{n+1}, \ldots\}$. (So, for example, $T_1 = \{x_1, x_2, x_3, \ldots\}$ and $T_2 = \{x_2, x_3, x_4, \ldots\}$.)

   Assume that T_1 has a lower bound. Deduce that for any n, the set T_n has a GLB, and call it b_n. Prove that $b_1 \leq b_2 \leq b_3 \leq \cdots$.

   For the following sequences $x_1, x_2, \ldots$, work out b_n, and also work out the LUB of the set $\{b_1, b_2, \ldots\}$ when it exists:

   (a) $x_1 = 1, x_2 = 2, x_3 = 3$, and in general $x_n = n$,

   (b) $x_1 = 1, x_2 = \frac{1}{2}, x_3 = \frac{1}{3}$, and in general $x_n = \frac{1}{n}$,

   (c) $x_1 = 1, x_2 = 2, x_3 = 1, x_4 = 2, x_5 = 1$, and so on, alternating between 1 and 2.

10. Critic Ivor Smallbrain reckons he has managed to prove that if n is any integer with $n \geq 3$, then there do not exist any positive integers a, b, c satisfying the equation $a^n + b^n = c^n$. He modestly calls this result "Smallbrain's first theorem" and attempts to write down a proof in the margin of his theatre programme during a particularly tedious performance of the Hungarian classic *Bevezetés — hoz analízis*. He fails to do so because the margin is too small.

    Now let us define a sequence $x_1, x_2, x_3, \ldots$ by letting $x_n = 1$ if there exist positive integers a, b, c satisfying the equation $a^n + b^n = c^n$, and letting $x_n = 2$ otherwise. Assuming that Smallbrain's first theorem is true, find the numbers b_n defined in Exercise 9, and the LUB of the set $\{b_1, b_2, \ldots\}$.

Chapter 23

More Analysis: Limits

Remember our discussion in Chapter 3 of how every real number has a decimal expression? We said that the expression $b_0.b_1b_2b_3\ldots$ represents the real number that is the "sum to infinity" of the series

$$b_0 + \frac{b_1}{10} + \frac{b_2}{10^2} + \frac{b_3}{10^3} + \cdots$$

In this chapter we aim to make this statement precise. To do so, we need to introduce one of the most fundamental concepts in mathematics — that of the *limit* of a sequence of real numbers.

First we need to say exactly what we mean by a sequence. That's easy enough: a sequence is just an infinite list $a_1, a_2, a_3, \ldots, a_n, \ldots$ of real numbers in a definite order. The number a_n is called the nth term of the sequence. We usually denote such a sequence just by the symbol (a_n).

Example 23.1

1. $1, 1, 1, \ldots$ is the sequence (a_n) where $a_n = 1$ for all n.

2. $1, \frac{1}{2}, \frac{1}{3}, \frac{1}{4}, \ldots$ is the sequence (a_n) with $a_n = \frac{1}{n}$.

3. If $a_n = (-1)^n$, the sequence (a_n) is $-1, 1, -1, 1, \ldots$

4. Define a_n to be 10^{-6} if n is prime, and $\frac{1}{n}$ if n is not prime. Then the sequence (a_n) is

$$1, 10^{-6}, 10^{-6}, \frac{1}{4}, 10^{-6}, \frac{1}{6}, 10^{-6}, \frac{1}{8}, \frac{1}{9}, \frac{1}{10}, 10^{-6}, \ldots$$

We now want to define the limit of a sequence (a_n). Intuitively, as discussed in Chapter 3, this should be a real number a such that we can make all the a_n's as close as we like to a provided we go far enough along the sequence.

How do we make this mathematically precise? To say that we can make the a_n's "as close as we like" to a means the following: if we pick *any* positive real number ε, however small, all the a_n's beyond a certain point lie between $a - \varepsilon$ and $a + \varepsilon$. So here now is the formal definition.

DEFINITION   *We say that a sequence (a_n) has a* limit *a if the following is true. Given any real number $\varepsilon > 0$, there is an integer N (which depends on ε) such that all the terms a_n for $n \geq N$ lie between $a - \varepsilon$ and $a + \varepsilon$.*

The definition is often written a little more succinctly, as follows: for any $\varepsilon > 0$, there exists N such that $|a_n - a| < \varepsilon$ for all $n \geq N$.

If the sequence (a_n) has the limit a, we write $a = \lim a_n$ or $a_n \to a$, and say that a_n *tends to the limit* a.

One might ask whether a sequence can have more than one limit. The answer is no: the limit of a sequence is unique (see Exercise 2 at the end of the chapter).

Example 23.2

Let's examine the first two sequences in Example 23.1. The first sequence is $1, 1, 1, \ldots$ and it is hard to imagine the limit of this sequence being anything but 1. Indeed it is 1. To see this we apply the definition. Let $\varepsilon > 0$ be any positive real number. Then, taking $N = 1$, we have $|a_n - 1| = 0 < \varepsilon$ for all $n \geq N$. So, according to the definition, $\lim a_n = 1$ as expected.

Now consider the second sequence, (a_n) with $a_n = \frac{1}{n}$. We show that the limit of this sequence is 0. Let $\varepsilon > 0$. Choose an integer $N > \frac{1}{\varepsilon}$. Then, for $n \geq N$,

$$|a_n - 0| = \frac{1}{n} \leq \frac{1}{N} < \varepsilon.$$

Hence $a_n \to 0$.

The order of events in the definition of a limit is very important. First we are given an arbitrary real number $\varepsilon > 0$; then we find an N that works, and, as in the second example above, N usually depends on ε. (Reversing the order of events and writing that there exists N such that for any $\varepsilon > 0$, $|a_n - a| < \varepsilon$ for all $n \geq N$, has quite a different meaning — see Exercise 5.)

Example 23.3

Now let's examine the third and fourth sequences in Example 23.1. The third sequence, (a_n) with $a_n = (-1)^n$, does not have a limit. Here is a proof of this by contradiction. Suppose (a_n) does have a limit. Call it a. To obtain a contradiction we need to find a specific value of $\varepsilon > 0$ for which there is *no* N such that $|a_n - a| < \varepsilon$ for all $n \geq N$. Let $\varepsilon = \frac{1}{4}$.

Then whatever a is, either 1 does not lie between $a - \varepsilon$ and $a + \varepsilon$, or -1 does not. So whatever N we try, there will be a value of $n \geq N$ such that $|a_n - a| > \varepsilon$. Hence there is no N which works in the definition of a limit for $\varepsilon = \frac{1}{4}$. This contradiction shows that (a_n) does not have a limit.

The fourth sequence in Example 23.1 also does not have a limit. We prove this in a similar way. Suppose that $a_n \to a$. If $a \neq 10^{-6}$, let $\varepsilon = |10^{-6} - a| > 0$. As $a_n \to a$, there exists N such that $|a_n - a| < \varepsilon$ for all $n \geq N$. However, there is a prime $p > N$ by Euclid's Theorem 12.1, and $a_p = 10^{-6}$. Then $|a_p - a| = |10^{-6} - a|$ is not less than ε, which is a contradiction. Hence a must be 10^{-6}. Now take $\varepsilon = \frac{10^{-6}}{2}$. As $a_n \to a = 10^{-6}$, there exists N such that $|a_n - 10^{-6}| < \varepsilon$ for all $n \geq N$. But for any non-prime value of n greater than 2×10^6 we have $a_n = \frac{1}{n} < \frac{10^{-6}}{2}$, hence $|a_n - 10^{-6}| > \frac{10^{-6}}{2} = \varepsilon$. This is a contradiction. Hence (a_n) does not have a limit.

The above examples show that some sequences have a limit and others don't. We call a sequence that has a limit a *convergent* sequence. In Example 23.1, the first two sequences are convergent and the third and fourth are not.

Here is another example which is a little trickier than the previous ones.

Example 23.4

Let (a_n) be the sequence with $a_n = \frac{3n^2 + 2n + 1}{n^2 - n - 3}$. Is (a_n) convergent, and if so, what is its limit?

Answer If we write a_n as $3 + \frac{5n + 10}{n^2 - n - 3}$, it seems plausible that (a_n) should be convergent and have the limit 3. Let's try to prove this.

Let $\varepsilon > 0$. We want to choose N such that for $n \geq N$,

$$|a_n - 3| = |\frac{5n + 10}{n^2 - n - 3}| < \varepsilon.$$

It looks awkward to calculate what N should be. But we can simplify things by observing that provided $n > 10$,

$$|\frac{5n + 10}{n^2 - n - 3}| < \frac{6n}{\frac{1}{2}n^2} = \frac{12}{n}.$$

Hence if we choose N to be greater than both 10 and $\frac{12}{\varepsilon}$, then for any $n \geq N$,

$$|a_n - 3| = |\frac{5n + 10}{n^2 - n - 3}| < \frac{12}{n} < \varepsilon.$$

This proves that the limit of the sequence (a_n) is 3.

Bounded Sequences

We say that a sequence (a_n) is *bounded* if it has both an upper and a lower bound — in other words, if there are real numbers L and U such that

$$L \leq a_n \leq U \quad \text{for all } n.$$

Notice that if we take $K = \max\{|L|, |U|\}$, we also have $|a_n| \leq K$ for all n.

For example, if $a_n = (-1)^n \sin n$, then the sequence (a_n) is bounded (since $|a_n| \leq 1$ for all n). On the other hand, the sequence (b_n) with $b_n = (-1)^n \sqrt{n}$ is not bounded.

PROPOSITION 23.1
Every convergent sequence is bounded.

PROOF   Let (a_n) be a convergent sequence, and let $a = \lim a_n$. Then, letting $\varepsilon = 1$, there exists N such that $|a_n - a| < 1$ for all $n \geq N$; in other words,

$$a - 1 < a_n < a + 1 \quad \text{for all } n \geq N.$$

Now put U equal to the maximum of the $N+1$ numbers $a_1, \ldots, a_N$ and $a+1$. Then $a_n \leq U$ for all n. Similarly, $a_n \geq L$ for all n, where L is the minimum of the numbers $a_1, \ldots, a_N$ and $a - 1$.   ∎

For example, the sequence (b_n) with $b_n = (-1)^n \sqrt{n}$ is not bounded, and hence is not convergent by Proposition 23.1.

Calculating Limits

Calculating limits from first principles just using the definition can be very awkward if the expression for a_n is not particularly simple. Fortunately, there are some basic aids to such calculations, provided by the following proposition.

PROPOSITION 23.2
Let (a_n) and (b_n) be convergent sequences, and suppose $a_n \to a$ and $b_n \to b$. Then the following hold.

(1) $a_n + b_n \to a + b$ (sum rule)

(2) $a_n b_n \to ab$ (*product rule*)

(3) If $b_n \neq 0$ *for all* n, *and* $b \neq 0$, *then* $\frac{a_n}{b_n} \to \frac{a}{b}$ (*quotient rule*).

PROOF   (1) Let $\varepsilon > 0$. Then also $\frac{\varepsilon}{2} > 0$, so there exist integers N and M such that

$$|a_n - a| < \tfrac{\varepsilon}{2} \text{ for } n \geq N,$$
$$|b_n - b| < \tfrac{\varepsilon}{2} \text{ for } n \geq M.$$

Let $R = \max(N, M)$. Then, for $n \geq R$,

$$|(a_n + b_n) - (a + b)| = |(a_n - a) + (b_n - b)|$$
$$\leq |a_n - a| + |b_n - b| \text{ (by the Triangle Inequality 4.12)}$$
$$< \tfrac{\varepsilon}{2} + \tfrac{\varepsilon}{2} = \varepsilon.$$

Hence $a_n + b_n \to a + b$.

(2) To show that $a_n b_n \to ab$, we will need to study the differences $a_n b_n - ab$, which we can cunningly rewrite as

$$a_n b_n - ab = (a_n - a) b_n + a(b_n - b).$$

By the Triangle Inequality 4.12,

$$|a_n b_n - ab| \leq |a_n - a| |b_n| + |a| |b_n - b|. \tag{23.1}$$

Now let $\varepsilon > 0$. We want to make the left-hand side of (23.1) less than ε, so we try to make each term on the right-hand side less than $\frac{\varepsilon}{2}$. Here's how.

Since (b_n) is convergent, it is bounded by Proposition 23.1, so there is a positive real number K such that $|b_n| \leq K$ for all n. Now $\frac{\varepsilon}{2K} > 0$, so there exists N such that

$$|a_n - a| < \frac{\varepsilon}{2K} \text{ for all } n \geq N.$$

Also $\frac{\varepsilon}{2(|a|+1)} > 0$, so there exists M such that

$$|b_n - b| < \frac{\varepsilon}{2(|a| + 1)} \text{ for all } n \geq M.$$

Let $R = \max(N, M)$. Then, for $n \geq R$, we see from (23.1) that

$$|a_n b_n - ab| < \frac{\varepsilon}{2K} \cdot K + |a| \cdot \frac{\varepsilon}{2(|a| + 1)} \leq \varepsilon.$$

It follows that $a_n b_n \to ab$.

(3) It is enough to show that $\frac{1}{b_n} \to \frac{1}{b}$, since it will follow from this that $\frac{a_n}{b_n} \to \frac{a}{b}$ by applying part (2) to the two sequences (a_n) and $(\frac{1}{b_n})$.

To show that $\frac{1}{b_n} \to \frac{1}{b}$, we will need to study

$$\left|\frac{1}{b_n} - \frac{1}{b}\right| = \frac{|b_n - b|}{|b_n||b|}. \tag{23.2}$$

To make the right-hand side of (23.2) small, we first need to get a lower bound for $|b_n|$. Now $\frac{1}{2}|b| > 0$, so there exists N such that

$$|b_n - b| < \frac{1}{2}|b| \quad \text{for all } n \geq N.$$

Since $|b_n - b| \geq |b| - |b_n|$ by Example 4.13, it follows that $|b| - |b_n| < \frac{1}{2}|b|$, and hence $|b_n| > \frac{1}{2}|b|$ for all $n \geq N$. Therefore, we now have the following upper bound for the right-hand side of (23.2):

$$\frac{|b_n - b|}{|b_n||b|} < \frac{|b_n - b|}{\frac{1}{2}|b|^2} \quad \text{for all } n \geq N. \tag{23.3}$$

Now let $\varepsilon > 0$. We want to make the right-hand side of (23.3) less than ε. Well, $\frac{1}{2}\varepsilon|b|^2 > 0$, so there exists M such that $|b_n - b| < \frac{1}{2}\varepsilon|b|^2$ for all $n \geq M$. Let $R = \max(N, M)$. Then for $n \geq R$, by (23.3),

$$\frac{|b_n - b|}{|b_n||b|} < \frac{\frac{1}{2}\varepsilon|b|^2}{\frac{1}{2}|b|^2} = \varepsilon,$$

and hence by (23.2), $\left|\frac{1}{b_n} - \frac{1}{b}\right| < \varepsilon$. It follows that $\frac{1}{b_n} \to \frac{1}{b}$, as required.

∎

Example 23.5

Find $\lim a_n$, where $a_n = \frac{3n^4 - 17n^2 + 12}{2n^4 + 5n^3 - 5}$.

Answer We would like to apply the quotient rule in Proposition 23.2(3), but we can't do this directly as neither the numerator nor the denominator in the expression for a_n is convergent. However, if we divide top and bottom by n^4, we can cleverly rewrite a_n as

$$a_n = \frac{3 - \frac{17}{n^2} + \frac{12}{n^4}}{2 + \frac{5}{n} - \frac{5}{n^4}}.$$

We know that $\frac{1}{n} \to 0$ by Example 23.2. So, by the product rule, $\frac{1}{n^2} \to 0$, $\frac{1}{n^3} \to 0$ and so on, and so using the sum rule we have

$$3 - \frac{17}{n^2} + \frac{12}{n^4} \to 3, \quad \text{and } 2 + \frac{5}{n} - \frac{5}{n^4} \to 2.$$

Hence, by the quotient rule, $a_n \to \frac{3}{2}$.

Increasing and Decreasing Sequences

DEFINITION   *A sequence (a_n) is increasing if $a_{n+1} \geq a_n$ for all n. Similarly, (a_n) is decreasing if $a_{n+1} \leq a_n$ for all n.*

For example, the sequences (a_n), (b_n) with $a_n = n$, $b_n = 1 - \frac{1}{n}$ are both increasing; the sequence (c_n) with $c_n = 1$ for all n is both increasing and decreasing; and the sequence (d_n) with $d_n = \frac{(-1)^n}{n}$ is neither increasing nor decreasing.
Here is a striking result on increasing sequences.

PROPOSITION 23.3
Let (a_n) be an increasing sequence which is bounded. Then (a_n) is convergent.

PROOF   Since (a_n) is bounded, the set $S = \{a_n : n \in \mathbb{N}\}$ has an upper bound. Hence, by the Completeness Axiom for $\mathbb{R}$ (see Chapter 22), S has a least upper bound — call it l. We shall prove that l is the limit of the sequence (a_n).

Let $\varepsilon > 0$. Then $l - \varepsilon$ is not an upper bound for the set S, so there exists N such that $a_N > l - \varepsilon$. As (a_n) is increasing, this implies that $a_n \geq a_N > l - \varepsilon$ for all $n \geq N$. Also l is an upper bound for S, so $a_n \leq l$ for all n. We conclude that $l - \varepsilon < a_n \leq l$ for all $n \geq N$, which means that $|a_n - l| < \varepsilon$ for all $n \geq N$. This shows that $a_n \to l$.   ∎

Similarly, any decreasing sequence (b_n) that is bounded must have a limit (see Exercise 6).

Proposition 23.3 has great significance for us. It means that we can finally put our theory of decimal expressions for real numbers on a rigorous footing. Let $b_0.b_1b_2b_3\ldots$ be a decimal expression with $b_0 \geq 0$. Define a sequence (a_n) as follows:

$$a_1 = b_0, \quad a_2 = b_0 + \frac{b_1}{10}, \quad a_3 = b_0 + \frac{b_1}{10} + \frac{b_2}{10^2},$$

and in general

$$a_n = b_0 + \frac{b_1}{10} + \cdots + \frac{b_{n-1}}{10^{n-1}}.$$

Then the sequence (a_n) is increasing, since $a_{n+1} - a_n = \frac{b_n}{10^n} \geq 0$. It is also bounded, since $a_n \geq b_0$ and also

$$a_n = b_0 + \frac{b_1}{10} + \cdots + \frac{b_{n-1}}{10^{n-1}} \leq b_0 + \frac{9}{10} + \cdots + \frac{9}{10^{n-1}} < b_0 + 1.$$

Hence, by Proposition 23.3, the sequence (a_n) has a limit l. So when we write

$$l = b_0.b_1 b_2 b_3 \ldots,$$

we mean precisely that l is the real number that is the limit of the above sequence (a_n).

Example 23.6
Define a sequence (e_n) as follows:

$$e_n = 1 + \frac{1}{1!} + \frac{1}{2!} + \cdots + \frac{1}{n!}.$$

So $e_1 = 2$, $e_2 = \frac{5}{2}$, $e_3 = \frac{8}{3}$ and so on. We saw in Example 8.4 that $2^n < n!$ for all $n \geq 4$, and it follows easily from this that $2^{n-1} \leq n!$ for all $n \geq 1$. Hence

$$e_n \leq 1 + 1 + \frac{1}{2} + \frac{1}{2^2} + \cdots + \frac{1}{2^{n-1}} = 1 + \frac{1 - (\frac{1}{2})^n}{1 - \frac{1}{2}} = 1 + 2\left(1 - \frac{1}{2^n}\right) < 3.$$

Thus (e_n) is an increasing sequence which is bounded, so by Proposition 23.3 it is convergent. We denote the limit of this sequence by the symbol e. This is a very famous real number, often called the "base of natural logarithms." You have probably come across it already in your studies.

As a decimal, $e = 2.71828\ldots$. A natural question arises: is e a rational number? The answer is to be found in Exercise 8 below.

Exercises for Chapter 23

1. Which of the following sequences (a_n) are convergent and which are not? For the convergent sequences, find the limit.

(i) $a_n = \frac{n}{n+5}$.

(ii) $a_n = \frac{1}{\sqrt{n+5}}$.

(iii) $a_n = \frac{n\sqrt{n}}{n+5}$.

(iv) $a_n = \frac{(-1)^n \sin n}{\sqrt{n}}$.

(v) $a_n = \frac{n^3 - 2\sqrt{n} + 7}{2 - n^2 - 5n^3}$.

(vi) $a_n = \frac{1 - (-1)^n n}{n}$.

(vii) $a_n = \sqrt{n+1} - \sqrt{n}$.

2. Prove that the limit of a sequence, if it exists, is unique: in other words, if (a_n) is a sequence such that $a_n \to a$ and $a_n \to b$, then $a = b$.

3. Let S be a non-empty set of real numbers, and suppose S has least upper bound c. Prove that there exists a sequence (s_n) such that $s_n \in S$ for all n and $s_n \to c$.

4. For each of the following sequences (a_n), decide whether it is (a) bounded, (b) increasing, (c) decreasing, (d) convergent:

(i) $a_n = \frac{n^3}{n^3-1}$.

(ii) $a_n = 2^{1/n}$.

(iii) $a_n = 1 - \frac{(-1)^n}{n}$.

(iv) $a_n = |5n - n^2|$.

5. If we reverse the order of events in the definition of the limit of a sequence (a_n), we get

$$\exists N \text{ such that } \forall \varepsilon > 0 \forall n \geq N, \ |a_n - a| < \varepsilon.$$

What does this mean?

6. Show that a bounded decreasing sequence is convergent.

7. A sequence (a_n) is defined by

$$a_1 = 1 \text{ and } a_{n+1} = \frac{a_n^2 + 2}{2a_n} \ \forall n \geq 1.$$

(i) Prove that (a_n) is a bounded sequence, and decreases for $n \geq 2$.

(ii) Show that the limit of (a_n) is $\sqrt{2}$.

8. This exercise contains a proof that the number e, defined in Example 23.6 as the limit of the sequence (e_n), is irrational.

(a) Show that $e_n = \frac{p_n}{n!}$, where p_n is an integer.

(b) Show that $e - e_n = \frac{1}{n!}(\frac{1}{n+1} + \frac{1}{(n+1)(n+2)} + \cdots)$. By comparing the term in brackets with a suitable geometric series, deduce that for all n,

$$0 < e - e_n < \frac{1}{n \cdot n!}.$$

(c) Deduce that $0 < n!e - p_n < \frac{1}{n}$ for all n.

(d) Now assume that e is rational. Show that $n!e \in \mathbb{Z}$ for some n. Hence obtain a contradiction.

9. Critic Ivor Smallbrain is sitting by the fire in his favourite pub, *The Fox and Bounds*. Also there are his friends Polly Gnomialle, Greta Picture, Gerry O'Laughing, Einstein, Hawking and celebrity mathematician Richard Thomas. Also joining them are great film directors Michael Loser and Ally Wooden.

All nine of them think that infinity is cool, and having seen the definition of what it means for a sequence (a_n) to have a limit a, they are trying to define what it should mean to say that $\lim a_n = \infty$. They decide that informally this should mean that you can make all the a_n's as large as you like, provided you go far enough along the sequence. They then take it in turns to try to write down a proper rigorous definition. Here are their attempts:

Michael Loser writes: $\forall a \in \mathbb{R}, a_n \not\to a$.

Polly writes: $\forall \varepsilon > 0 \; \exists N \in \mathbb{N}$ such that $n \geq N \Rightarrow |a_n - \infty| < \varepsilon$.

Greta writes: $\forall R > 0 \; \exists N \in \mathbb{N}$ such that $n \geq N \Rightarrow a_n > R$.

Ally Wooden writes: $\forall l \in \mathbb{R} \; \forall \varepsilon \in \mathbb{R} \; \exists N \in \mathbb{N}$ such that $n \geq N \Rightarrow |a_n - l| > \varepsilon$.

Gerry writes: $\forall a \in \mathbb{R} \; \exists \varepsilon > 0$ such that $\forall N \in \mathbb{N} \; \exists n \geq N$ such that $|a_n - a| < \varepsilon$.

Einstein writes: $\forall \varepsilon > 0 \; \exists N \in \mathbb{N}$ such that $\forall n \geq N, a_n > \frac{1}{\varepsilon}$.

Hawking writes: $\forall n \in \mathbb{N}, a_{n+1} > a_n$.

Richard Thomas writes: $\exists N \in \mathbb{N}$ such that $\forall R > 0, \forall n \geq N, a_n > R$

Ivor writes: $\forall R \in \mathbb{R} \; \exists n \in \mathbb{N}$ such that $a_n > R$.

Who do you think is right and who do you think is wrong? (There may be more than one who is right!) For the wrong ones, illustrate why you think they are wrong with an example.

Chapter 24

Yet More Analysis: Continuity

In this chapter I am going to introduce you to the notion of a continuous function of the real numbers. This is a topic that deserves a whole book to itself, but I am giving you this little taster here for several reasons. First, this notion is the right setting in which to prove the existence of nth roots of positive real numbers, which I promised to do way back in Chapter 5. Second, one can quite quickly develop enough theory to prove a famous result called the Intermediate Value Theorem, which sheds quite a bit of light on the great Fundamental Theorem of Algebra, stated in Chapter 7. And last, the idea of a continuous function provides yet another example of something natural and obvious-sounding being turned into rigorous mathematics, with fruitful results.

The material in this chapter is perhaps at a higher level than most of the rest of the book, but it will repay study and hopefully serve as a useful introduction to your next course in analysis.

Continuous Functions

Let f be a function from $\mathbb{R}$ to $\mathbb{R}$, and let c be a real number. We say that f is *continuous* at c if we can make $f(x)$ as close as we like to $f(c)$, provided we confine x to being in a sufficiently small neighbourhood of c. We shall make the phrases "as close as we like" and "sufficiently small neighbourhood" precise in a rather similar fashion to the definition of a limit in the previous chapter. Here is the formal definition.

DEFINITION   *A function $f : \mathbb{R} \to \mathbb{R}$ is continuous at a point $c \in \mathbb{R}$ if the following holds. For any $\varepsilon > 0$, there exists $\delta > 0$ (depending on ε) such that*

$$|x - c| < \delta \Rightarrow |f(x) - f(c)| < \varepsilon.$$

In other words, $f(x)$ can be confined between $f(c) - \varepsilon$ and $f(c) + \varepsilon$ provided x is confined between $c - \delta$ and $c + \delta$.

Some examples may serve to make the definition clearer.

Example 24.1

1. The function $f(x) = x$ is continuous at every point $c \in \mathbb{R}$. To see this, let $\varepsilon > 0$. Take $\delta = \varepsilon$. Then

$$|x - c| < \delta \Rightarrow |f(x) - f(c)| = |x - c| < \delta = \varepsilon.$$

2. Any constant function $f(x) = k$ for all $x \in \mathbb{R}$ is continuous at every point in $\mathbb{R}$ (Exercise 1 at the end of the chapter).

3. Define $f : \mathbb{R} \to \mathbb{R}$ by

$$f(x) = \begin{cases} 1, & \text{if } x \in \mathbb{Q} \\ 0, & \text{if } x \notin \mathbb{Q}. \end{cases}$$

We claim that f is *not* continuous at any point $c \in \mathbb{R}$. To prove this, we must produce a value of $\varepsilon > 0$ such that there is no δ satisfying the definition.

First let $c \in \mathbb{Q}$, so $f(c) = 1$. Take $\varepsilon = \frac{1}{2}$. For any $\delta > 0$, we know by Proposition 2.5 that there is an irrational number r between $c - \delta$ and $c + \delta$. Then $|r - c| < \delta$, but $|f(r) - f(c)| = |0 - 1| = 1 > \varepsilon$. So there is no value of δ which works for $\varepsilon = \frac{1}{2}$, showing that f is not continuous at c.

Now let $c \notin \mathbb{Q}$, so $f(c) = 0$. Again take $\varepsilon = \frac{1}{2}$. This time we observe that for any $\delta > 0$, there is a rational number s between $c - \delta$ and $c + \delta$. Then $|s - c| < \delta$ but $|f(s) - f(c)| = 1 > \varepsilon$, showing that f is not continuous at c.

We say that a function $f : \mathbb{R} \to \mathbb{R}$ is *continuous on* $\mathbb{R}$ if it is continuous at every point $c \in \mathbb{R}$. Naively, this says that the graph of $y = f(x)$ is a continuous line which can be drawn without taking your pen off the paper (but this is a bit misleading, as there are continuous functions whose graphs cannot even be drawn).

From Example 24.1 we see that constant functions are continuous on $\mathbb{R}$, and also the function $f(x) = x$ is continuous on $\mathbb{R}$. Having obtained these simple examples, we can build up many more using the following sum, product and quotient rules.

PROPOSITION 24.1

Let f and g be functions from $\mathbb{R}$ to $\mathbb{R}$.

(1) *(Sum rule): If f and g are continuous at a point c, then the sum $f + g$ is continuous at c.*

(2) *(Product rule): If f and g are continuous at c, then the product function f.g (defined by $(f.g)(x) = f(x)g(x)$) is continuous at c.*

(3) *(Quotient rule): If f and g are continuous at c, and $g(x) \neq 0$ for all x, then the quotient $\frac{f}{g}$ is continuous at c.*

PROOF   This is quite similar to the proof of the sum, product and quotient rules for limits in Proposition 23.2. We'll just prove the product rule (2) and leave the other parts to Exercise 2.

For convenience, write

$$A = f(c),\ B = g(c).$$

We need to study the difference $f(x)g(x) - AB$, which we rewrite as

$$f(x)g(x) - AB = (f(x) - A)g(x) + (g(x) - B)A.$$

From the Triangle Inequality 5.12, this implies

$$|f(x)g(x) - AB| \leq |f(x) - A|\,|g(x)| + |g(x) - B|\,|A|. \qquad (24.1)$$

To make the right-hand side small, we first need to bound $|g(x)|$. Now g is continuous at c, so (taking $\varepsilon = 1$ in the definition) there exists $\delta_0 > 0$ such that

$$|x - c| < \delta_0 \Rightarrow |g(x) - B| < 1$$
$$\Rightarrow |g(x)| < 1 + |B|.$$

Now let $\varepsilon > 0$. Define

$$\varepsilon_1 = \frac{\varepsilon}{2(1 + |B|)}, \quad \varepsilon_2 = \frac{\varepsilon}{2(1 + |A|)}.$$

Then $\varepsilon_1, \varepsilon_2 > 0$, so there exist $\delta_1, \delta_2 > 0$ such that

$$|x - c| < \delta_1 \Rightarrow |f(x) - A| < \varepsilon_1 \quad \text{and} \quad |x - c| < \delta_2 \Rightarrow |g(x) - B| < \varepsilon_2.$$

Let $\delta = \min(\delta_0, \delta_1, \delta_2)$. Then, by (24.1),

$$|x - c| < \delta \Rightarrow |f(x)g(x) - AB| \leq |f(x) - A|\,|g(x)| + |g(x) - B|\,|A|$$
$$< \varepsilon_1(1 + |B|) + \varepsilon_2|A|$$
$$= \frac{\varepsilon}{2(1+|B|)}(1 + |B|) + \frac{\varepsilon}{2(1+|A|)}|A|$$
$$< \varepsilon.$$

Hence $f.g$ is continuous at c.   ∎

Example 24.2

1. We know that the function $f(x) = x$ is continuous on $\mathbb{R}$. By the

product rule in Proposition 24.1(2), the function $x \to x^2$ is also continuous, as is $x \to x^3$, and in general $x \to x^n$ for any positive integer n. Also constant functions are continuous, so the product rule also shows that for any real number c the function $x \to cx^n$ is continuous. Hence, using the sum rule, we see that any polynomial function

$$x \to a_n x^n + a_{n-1} x^{n-1} + \cdots + a_1 x + a_0 \quad (a_i \in \mathbb{R})$$

is continuous on $\mathbb{R}$.

2. Consider the function

$$g(x) = \frac{x^3 - 5x^2 + 2}{x^2 + 2x + 2}.$$

The denominator $x^2 + 2x + 2 = (x+1)^2 + 1 > 0$ for all x, so g is continuous on $\mathbb{R}$ by the quotient rule.

3. Now consider the function

$$h(x) = \begin{cases} \frac{1}{x}, & \text{if } x \neq 0 \\ 0, & \text{if } x = 0. \end{cases}$$

A slight generalisation of the quotient rule shows that h is continuous at any point $c \neq 0$. But is h continuous at 0? The answer is no. To see this, let $\varepsilon = \frac{1}{2}$, say. Whatever $\delta > 0$ we choose, there exists a value of x with $|x - 0| < \delta$ such that $|h(x) - h(0)| > \varepsilon$ (any x with $|x| < 2$ will do). Hence h is not continuous at 0.

The Intermediate Value Theorem

This famous theorem says that if f is a continuous function and a, b are real numbers such that $f(a) \neq f(b)$, then f takes on all values between $f(a)$ and $f(b)$. If we think of a continuous function as one whose graph is a continuous line, this seems obvious, as a continuous line from $(a, f(a))$ to $(b, f(b))$ will surely pass through points with all possible y-values between $f(a)$ and $f(b)$. However, finding a rigorous proof is not at all obvious, and the following proof is one of the trickiest in the book.

THEOREM 24.1 Intermediate Value Theorem
Let $f : \mathbb{R} \to \mathbb{R}$ be a continuous function on $\mathbb{R}$, and let $a, b \in \mathbb{R}$ with $a < b$ and $f(a) \neq f(b)$. Then, for any real number γ between $f(a)$ and $f(b)$, there exists a real number c between a and b such that $f(c) = \gamma$.

PROOF   Let's assume that $f(a) < f(b)$. (The case where $f(a) > f(b)$ will follow by a simple trick which we give at the end of the proof.) So we have

$$f(a) < \gamma < f(b).$$

Define a set S of real numbers as follows:

$$S = \{x \in \mathbb{R} : x \le b \text{ and } f(x) < \gamma\}.$$

Then S is non-empty as $a \in S$, and S has b as an upper bound. Therefore, by the Completeness Axiom (see Chapter 22), S has a least upper bound: let

$$c = \mathrm{LUB}(S).$$

We shall prove that $f(c) = \gamma$, which will prove the theorem.

Suppose first that $f(c) < \gamma$. Let $\varepsilon = \gamma - f(c) > 0$. As f is continuous at c, there exists $\delta > 0$ such that

$$|x - c| < \delta \Rightarrow |f(x) - f(c)| < \varepsilon = \gamma - f(c).$$

Hence there is a value $x = c + \frac{\delta}{2}$, for instance, such that $f(c + \frac{\delta}{2})$ lies between $f(c) - \varepsilon$ and $f(c) + \varepsilon$, so that

$$f(c + \frac{\delta}{2}) < f(c) + \varepsilon = \gamma.$$

But this means that $c + \frac{\delta}{2} \in S$. This is a contradiction, since c is an upper bound for S.

Now suppose that $f(c) > \gamma$ and let $\varepsilon = f(c) - \gamma > 0$. As f is continuous at c, there exists $\delta > 0$ such that $|x - c| < \delta \Rightarrow |f(x) - f(c)| < \varepsilon$. So whenever $c - \delta < x \le c$, we have $f(x) > f(c) - \varepsilon = \gamma$. This means that none of the real numbers between $c - \delta$ and c lie in S, which is to say that $c - \delta$ is an upper bound for S. This is a contradiction, as c is the least upper bound of S.

We have shown that both of the assumptions $f(c) < \gamma$ and $f(c) > \gamma$ lead to contradictions, so we conclude that $f(c) = \gamma$, as required.

The case where $f(a) > f(b)$ can be deduced by considering the function $g(x) = -f(x)$: we have $g(a) < -\gamma < g(b)$, so from what we have already proved, there exists c between a and b such that $g(c) = -\gamma$, which means that $f(c) = \gamma$.   ∎

Existence of nth Roots

Our first application of the Intermediate Value Theorem is to prove the existence of nth roots of positive real numbers, stated long ago as Proposition 5.1.

PROPOSITION 24.2
Let n be a positive integer. If d is a positive real number, then there is exactly one positive real number c such that $c^n = d$.

PROOF   Define $f : \mathbb{R} \to \mathbb{R}$ to be the function

$$f(x) = x^n - d.$$

Then f is a polynomial function, and hence is continuous on $\mathbb{R}$ by Example 24.2.

Now $f(0) = -d < 0$. Also, if we choose b to be a real number such that $b > \max(1, d)$, then $f(b) = b^n - d > 0$. Hence

$$f(0) < 0 < f(b).$$

By the Intermediate Value Theorem, there is a real number c between 0 and b such that $f(c) = 0$, which is to say that $c^n - d = 0$, so $c^n = d$.

This proves the existence of c. The uniqueness is easy: suppose $c_1^n = d$ with $c_1 > 0$. If $c_1 > c$ then $c_1^n > c^n$, and if $c_1 < c$ then $c_1^n < c^n$. Neither of these is possible, so $c_1 = c$, proving the uniqueness of c.   ∎

A Special Case of the Fundamental Theorem of Algebra

Remember the Fundamental Theorem of Algebra, stated as Theorem 7.1: every polynomial equation of degree at least 1 has a root in $\mathbb{C}$. This is quite a sophisticated result, whose proof you will probably have to wait to see until you take a course on complex analysis. However, there is a really neat proof that works for real polynomials of *odd* degree and uses just the Intermediate Value Theorem. Here it is.

PROPOSITION 24.3
Let $p(x)$ be a polynomial in x of odd degree with real coefficients. Then the equation $p(x) = 0$ has a real root.

PROOF   Let $p(x) = a_n x^n + \cdots + a_1 x + a_0$ with $a_n \neq 0$ and n odd. We may as well assume that $a_n > 0$ (otherwise replace $p(x)$ by $-p(x)$, which has the same roots).

Now for large positive x, the value of $p(x)$ is dominated by the leading term $a_n x^n$, so there is a positive real number b such that $p(b) > 0$. (This is a bit hand-wavy, but the details are easily filled in. I leave this to you in Exercise 3.) Similarly, for large negative x the dominant term is also $a_n x^n$, which is now negative because n is odd, so there is a negative real number a such that $p(a) < 0$. Thus

$$p(a) < 0 < p(b).$$

Now $p(x)$ is a continuous function, so by the Intermediate Value Theorem there is a real number c between a and b such that $p(c) = 0$. Then c is a root of $p(x)$, as required. ∎

Exercises for Chapter 24

1. Prove that any constant function is continuous on $\mathbb{R}$.

2. Prove the sum and quotient rules in Proposition 24.1.

3. Let $p(x) = a_n x^n + \cdots + a_1 x + a_0$ with all $a_i \in \mathbb{R}$, $a_n \neq 0$ and n odd. Prove that there exist real numbers a, b such that $p(a) < 0$ and $p(b) > 0$.

4. Show that each of the following functions is continuous at 0:

   (i) $x \to x \sin x$.

   (ii) $x \to |x|$.

   (iii) $x \to \begin{cases} x, & \text{if } x \in \mathbb{Q} \\ -x, & \text{if } x \notin \mathbb{Q}. \end{cases}$

5. Prove that if $f : \mathbb{R} \to \mathbb{R}$ is continuous at a, then so is the function $x \to |f(x)|$.

6. Let $f : \mathbb{R} \to \mathbb{R}$ be continuous on $\mathbb{R}$. Suppose that $f(x) = 0$ for all $x \in \mathbb{Q}$. Prove that $f(x) = 0$ for all $x \in \mathbb{R}$.

7. Prove that if f is continuous on $\mathbb{R}$ and $f(0) = f(1)$, then there exists c such that $\frac{1}{2} \leq c \leq 1$ and $f(c) = f(c - \frac{1}{2})$.

8. Let $f : \mathbb{R} \to \mathbb{R}$ be a function that is continuous on $\mathbb{R}$ and satisfies $|f(x)| \leq 1$ for all $x \in \mathbb{R}$.

   (i) Prove that if $P(x)$ is any polynomial of odd degree with real coefficients, then the equation $P(x) = f(x)$ has at least one real solution for x.

   (ii) Find a polynomial $P(x)$ of even degree at least 2, such that the equation $P(x) = f(x)$ has no real solutions.

9. Let $g : \mathbb{R} \to \mathbb{R}$ be a function such that

$$g(x+y) = g(x)g(y) \quad \text{for all } x, y \in \mathbb{R}.$$

(i) Show that $g(0)$ is equal to 0 or 1.

(ii) Show that if g is continuous at 0, then g is continuous on $\mathbb{R}$.

(iii) Suppose $g(a) = 0$ for some $a \in \mathbb{R}$. Prove that $g(x) = 0$ for all $x \in \mathbb{R}$.

10. Critic Ivor Smallbrain is in search of a wife. He is torn between Polly Gnomialle and Greta Picture, so he decides to set a challenge. He asks Polly to define a function $f : \mathbb{R} \to \mathbb{R}$ that is continuous at every irrational number and is discontinuous at every rational number. (Discontinuous just means not continuous.) If Polly gets it right he will choose her; if not, he will choose Greta.

Here is the function Polly comes up with:

$$P(x) = \begin{cases} 1, & \text{if } x = 0 \\ \frac{1}{n}, & \text{if } x = \frac{m}{n} \text{ is a rational in lowest terms, with } n > 0 \\ 0, & \text{if } x \text{ is irrational.} \end{cases}$$

Assuming that Ivor is so irresistible that neither will turn him down, predict who will become Mrs. Smallbrain.

Chapter 25

Introduction to Abstract Algebra: Groups

In the final two chapters of the book — this one and the next — I am going to introduce you to the theory of groups. This theory forms part of a vast area known as "abstract algebra". In abstract algebra we study certain basic systems in which we have a set, together with rules for combining any two elements of the set to get another element of the set; these rules are subject to various clearly defined assumptions, called "axioms." In group theory, as we shall see, there is just one rule for combining elements, and there are four axioms. You should not think that these axioms were thought up by some clever person who one sunny day sat down and wrote them down. Rather, they emerged over a long period — many different cases of what have come to be known as groups were studied in the eighteenth and nineteenth centuries, but it was not until late in the nineteenth century that the notion of an abstract group was introduced.

Group theory is a huge subject, and actually has many applications in other parts of mathematics, and also in other sciences. Once again I am giving you a taster to let you have the flavour of a different kind of topic to the rest of the book. We shall also see quite a few links with material in previous chapters.

Definition and Examples of Groups

Let S be a set. A *binary operation* $*$ on S is a rule which assigns to any ordered pair (a, b) $(a, b \in S)$ an element $a * b \in S$. In other words, it is a function from $S \times S$ to S.

Example 25.1

Here are a few examples of binary operations.

(1) $S = \mathbb{Z}$, $a * b = a + b$ for all $a, b \in S$.

(2) $S = \mathbb{C}$, $a*b = ab$ for all $a,b \in S$.

(3) $S = \mathbb{R}$, $a*b = a-b$ for all $a,b \in S$.

(4) $S = \mathbb{R}$, $a*b = ab+a+b$ for all $a,b \in S$.

(5) Let $S = S_n$ (the set of all permutations of $\{1,\ldots,n\}$), and for $f,g \in S$ define $f*g = f \circ g$, the composition of the permutations f and g.

(6) Let $S = \{1,2,3\}$, and define $*$ as follows:

$$1*1 = 3,\ 1*2 = 1,\ 1*3 = 2,$$
$$2*1 = 1,\ 2*2 = 2,\ 2*3 = 3,$$
$$3*1 = 2,\ 3*2 = 3,\ 3*3 = 1.$$

Given a binary operation $*$ on a set S and $a,b,c \in S$, we can form "$a*b*c$" in two ways, namely $(a*b)*c$ or $a*(b*c)$. These may or may not be equal. For instance, in Example 25.1 (1), $(a*b)*c = a*(b*c)$ for all $a,b,c \in S$ (see Rules 2.1(2)). However in (3) this is not the case, since for example

$$(3*5)*4 = (3-5)-4 = -6,$$
$$3*(5*4) = 3-(5-4) = 2.$$

DEFINITION   *A binary operation $*$ on a set S is* associative *if $(a*b)*c = a*(b*c)$ for all $a,b,c \in S$.*

The binary operation in Example 25.1(1) is associative, and so is the one in (2) (see Exercise 1(e) of Chapter 6); so is the one in (5), as was shown in Proposition 20.2, but the operation in (3) is not. The associativity or non-associativity of the binary operations in Example 25.1(4) and (6) is not quite so obvious, and I leave you to ponder these as part of Exercises 1 and 2 at the end of the chapter.

We are now ready to define a group.

DEFINITION   *A* group $(G,*)$ *is a set G with a binary operation $*$ satisfying the following axioms:*

(1) $a*b \in G$ for all $a,b, \in G$ **(closure axiom)**;

(2) $(a*b)*c = a*(b*c)$ for all $a,b,c \in G$ **(associativity axiom)**;

(3) *there exists an element $e \in G$ such that $a*e = e*a = a$ for all $a \in G$* **(identity axiom)**;

(4) *for any $a \in G$ there exists an element $a' \in G$ such that $a*a' = a'*a = e$* **(inverse axiom)**.

The element e in axiom (3) is called an *identity element* of G, and the element a' in (4) is an *inverse* of a.

Strictly speaking, the closure axiom (1) is not needed, as it is part of the definition of a binary operation; but it is usually included as an axiom to remind us that we need to check it.

Example 25.2

(1) $(\mathbb{Z}, +)$ is a group: closure is clear, associativity has already been remarked on, an identity element is the integer 0 (since $a + 0 = 0 + a = a$ for all $a \in \mathbb{Z}$), and for $a \in \mathbb{Z}$, an inverse is $-a$. Similarly $(\mathbb{Q}, +)$, $(\mathbb{R}, +)$ and $(\mathbb{C}, +)$ are also groups.

(2) $(\mathbb{Z}, \times)$ is not a group: it satisfies the first three axioms, but not the inverse axiom. A little argument is needed to see this. The only integer e satisfying $a \times e = e \times a = a$ for all $a \in \mathbb{Z}$ is $e = 1$, so 1 is the only identity element. But then the integer 2, for example, has no inverse as there is no integer x such that $2 \times x = 1$.

(3) What about $(\mathbb{Q}, \times)$? Again the first three axioms hold (with identity element 1), and it seems that there is no longer any problem with the inverse axiom, since the inverse of the rational $\frac{m}{n}$ is $\frac{n}{m}$. However, this is not quite right: the rational number 0 has no inverse, so this is not a group.

(4) In the light of the previous example, define $\mathbb{Q}^* = \mathbb{Q} - \{0\}$, the set of nonzero rationals. Then $(\mathbb{Q}^*, \times)$ is a group. So are $(\mathbb{R}^*, \times)$ and $(\mathbb{C}^*, \times)$.

(5) Let G be the set of complex numbers $\{1, -1, i, -i\}$. Let's show that $(G, \times)$ is a group, where $\times$ is just complex multiplication. First we need to check the closure axiom. A good way to do this is to write down the multiplication table of G:

	1	−1	i	−i
1	1	−1	i	−i
−1	−1	1	−i	i
i	i	−i	−1	1
−i	−i	i	1	−1

Associativity for $(G, \times)$ follows from the associativity of multiplication of complex numbers. An identity element in G is 1, and the existence of inverses for each element can be seen from the table. Hence $(G, \times)$ is a group.

(6) In the same way (much more easily) we can see that $(\{1, -1\}, \times)$ is a group.

(7) Let n be a positive integer, and as in Chapter 20 let S_n be the set of all permutations of $\{1,\ldots,n\}$. Proposition 20.2 shows that $(S_n, \circ)$ is a group, where $\circ$ is the binary operation of composition of functions. This group is known as the *symmetric group of degree n*.

We say that a group $(G, *)$ is *finite* if $|G|$ (the number of elements in the set G) is finite; we call $|G|$ the *size* of G. And we say $(G, *)$ is an *infinite* group if G is an infinite set. The groups in Example 25.2(1) and (4) are infinite; the groups in (5) and (6) are finite, of sizes 4 and 2; and the symmetric group S_n is finite and has size $n!$ (see Proposition 20.1).

In the examples (1), (4), (5) and (6) above we have $a * b = b * a$ for all elements a, b in the group. But this is not the case for the symmetric group S_n in example (7) when $n \geq 3$; to see this, let $a = (1\,2\,3)$ and $b = (1\,2)$ (here I am not writing down the 1-cycles in a and b), and observe that

$$a \circ b = (1\,3), \quad b \circ a = (2\,3).$$

This is such an important distiction between groups that we give it a special name:

DEFINITION   *A group $(G, *)$ is* abelian *if $a * b = b * a$ for all $a, b \in G$.*

Thus the groups $(\mathbb{Z}, +)$, $(\mathbb{Q}^*, \times)$ and so on are abelian, whereas $(S_n, \circ)$ is non-abelian for $n \geq 3$.

First Results

It's time to begin seeing what can be deduced from the axioms of group theory. On the face of it, there is nothing in the axioms to say that a group can't have several different identity elements, or that an element can't have several different inverses. Our first result addresses these questions.

PROPOSITION 25.1
Let $(G, *)$ be a group. Then

(i) *G has exactly one identity element;*
(ii) *each element of G has exactly one inverse.*

PROOF   (i) Suppose e and f are identity elements of G. Then for all $x \in G$,

$$x * e = e * x = x \text{ and } x * f = f * x = x.$$

Taking $x = f$ in the left-hand equations gives $e * f = f$; and taking $x = e$ in the right-hand equations gives $e * f = e$. Hence $e = f$. This shows that there is only one identity element in G.

(ii) Let $a \in G$, and suppose that a' and a'' are both inverses of a. Then

$$a * a' = a' * a = e \text{ and } a * a'' = a'' * a = e.$$

We now cleverly use associativity to show that $a' = a''$:

$$\begin{aligned} a' &= a' * e \\ &= a' * (a * a'') \\ &= (a' * a) * a'' \\ &= e * a'' \\ &= a''. \end{aligned}$$

Thus $a' = a''$, which proves (ii). ■

Notation Usually we write e for *the* identity element of a group $(G, *)$, and a^{-1} for *the* inverse of an element a. So in the group $(\mathbb{Q}^*, \times)$, the inverse x^{-1} is the rational $\frac{1}{x}$, while in $(\mathbb{Z}, +)$ the inverse x^{-1} is $-x$. In the latter example the equation $x^{-1} = -x$ may seem a bit strange at first sight, but one soon gets used to such things.

Also, instead of the phrase "$(G, *)$ is a group" we often say "G is a group under $*$." For example, to say that $(\mathbb{Z}, +)$ is a group is the same as saying that $\mathbb{Z}$ is a group under addition; likewise, S_n is a group under composition of functions.

Here is one more simple general property of groups.

PROPOSITION 25.2
In any group $(G, *)$, the following "cancellation laws" hold for all $a, x, y \in G$:

(i) $a * x = a * y \Rightarrow x = y$
(ii) $x * a = y * a \Rightarrow x = y$.

PROOF For (i), observe that

$$\begin{aligned} a * x = a * y &\Rightarrow a^{-1} * (a * x) = a^{-1} * (a * y) \\ &\Rightarrow (a^{-1} * a) * x = (a^{-1} * a) * y \\ &\Rightarrow e * x = e * y \\ &\Rightarrow x = y. \end{aligned}$$

Part (ii) is similar. ■

Multiplicative Notation for Groups

From now on, if $(G, *)$ is a group we shall usually write just ab instead of $a * b$. In view of the associativity axiom, $(ab)c = a(bc)$, and we just write abc for this element. We define powers of an element $a \in G$ as follows;

$$a^0 = e,$$
$$a^1 = a,$$
$$a^2 = a * a,$$
$$a^3 = a * a * a = a^2 * a,$$
$$a^n = a^{n-1} * a$$

for $n > 0$, and

$$a^{-n} = a^{-1} * \cdots * a^{-1} = (a^{-1})^n.$$

For example, in the group $(\mathbb{R}^*, \times)$, for $a \in \mathbb{R}^*$ the power a^n is just the usual n^{th} power of the real number a; whereas in the group $(\mathbb{Z}, +)$, a^n is the integer na.

The next result looks obvious, but does need to be proved.

PROPOSITION 25.3
For any $a \in G$ and any $m, n \in \mathbb{Z}$,

$$a^m a^n = a^{m+n}.$$

PROOF   For $m, n > 0$ this is indeed obvious, since

$$a^m a^n = \underbrace{a \cdots a}_{m} \underbrace{a \cdots a}_{n} = a^{m+n}.$$

For $m \geq 0, n < 0$,

$$a^m a^n = \underbrace{a \cdots a}_{m} \underbrace{a^{-1} \cdots a^{-1}}_{-n} = a^{m-(-n)} = a^{m+n},$$

and similarly for $m < 0, n \geq 0$. Finally, when $m, n < 0$,

$$a^m a^n = \underbrace{a^{-1} \cdots a^{-1}}_{-m} \underbrace{a^{-1} \cdots a^{-1}}_{-n} = a^{-(-m-n)} = a^{m+n}.$$

In future, when we write "let G be a group," we mean that the binary operation $*$ for G is understood, and we are writing ab instead of $a * b$. With this understanding, for a finite group $G = \{a, b, c, \ldots\}$ we define the *group table* of G to be the "multiplication table":

$$
\begin{array}{c|cccc}
 & a & b & c & \cdots \\
\hline
a & a^2 & ab & ac & \cdots \\
b & ba & b^2 & bc & \cdots \\
\vdots & \vdots & & &
\end{array}
$$

The group table of $(\{1, -1, i, -i\}, \times)$ is given in Example 25.2(5).

Example 25.3

In this example we calculate the group table of the symmetric group S_3. We could of course list the 6 elements of this group as in Example 20.2, and work out their products (as was started in Example 20.3), but it is more instructive to use some more concise notation. Define

$$a = (1\,2\,3),\ b = (1\,2) \in S_3.$$

Check that $a^2 = (1\,3\,2)$, $ab = (1\,3)$ and $a^2b = (2\,3)$. Thus the 6 elements of S_3 are

$$e, a, a^2, b, ab, a^2b. \tag{25.1}$$

By the closure axiom, the product ba is an element of S_3 and so is equal to one of the elements in the above list. Which one? Well, $ba = (2\,3)$, so it is a^2b. Thus we have the following equations:

$$a^3 = e,\ \ b^2 = e,\ \ ba = a^2b.$$

Using these equations, we can easily work out all products among the elements in the list (25.1). For example, here are the steps to calculate the product of b and a^2b:

$$ba^2b = baab = a^2bab = a^2a^2bb = a^4b^2 = aa^3b^2 = aee = a.$$

Here is the group table.

$$
\begin{array}{c|cccccc}
 & e & a & a^2 & b & ab & a^2b \\
\hline
e & e & a & a^2 & b & ab & a^2b \\
a & a & a^2 & e & ab & a^2b & b \\
a^2 & a^2 & e & a & a^2b & b & ab \\
b & b & a^2b & ab & e & a^2 & a \\
ab & ab & b & a^2b & a & e & a^2 \\
a^2b & a^2b & ab & b & a^2 & a & e
\end{array}
$$

A slightly more complicated example along these lines can be found in Exercise 7 at the end of the chapter.

Exercises for Chapter 25

1. Which of the following sets S are groups, under the stated binary operations?

   (i) $S = \{z \in \mathbb{C} : |z| = 1\}$ under the usual complex multiplication.

   (ii) $S = \mathbb{R} - \{-1\}$ under the binary operation $a * b = ab + a + b$ for all $a, b, \in S$.

   (iii) Let S be the set of all 97 mathematicians in the Maths department at Imperial College, with binary operation

   $$a * b = \text{ the better mathematician out of } a \text{ and } b.$$

   (You can assume that Liebeck $* x = x$ for all $x \in S$.)

   (iv) $S = \{x \in \mathbb{R} : x \geq 0\}$ with binary operation $a * b = \max(a, b)$ (the maximum of a and b) for all $a, b \in S$.

   (v) $S = \{z \in \mathbb{C} : z^3 - z^2 + z - 1 = 0\}$ under the usual complex multiplication.

   (vi) $S = \mathbb{C} - \{0\}$, with binary operation $a * b = |a| b$ for all $a, b, \in S$.

   (vii) $S = $ the set of all rational numbers with odd denominators, under the usual addition of rationals.

   (viii) $S = \{a, b\}$, with the binary operation $*$ defined by

   $$a * a = a, \ b * b = b, \ a * b = b, \ b * a = b.$$

   (ix) $S = \{a, b\}$, with the binary operation $*$ defined by

   $$a * a = a, \ b * b = a, \ a * b = b, \ b * a = b.$$

2. (a) Let $G = \{z \in \mathbb{C} : z^3 = 1\}$. Show that G is a group under the usual complex multiplication, and write down the group table of G.

   (b) Let $S = \{1, 2, 3\}$, with binary operation $*$ defined as in Example 25.1(6). Is $(S, *)$ a group?

3. Let G be a group, and let $a, b \in G$. Prove that
   (i) $(a^{-1})^{-1} = a$, and
   (ii) $(ab)^{-1} = b^{-1}a^{-1}$.

4. (a) Show that if G is an abelian group and n is an integer, then $(ab)^n = a^n b^n$ for all $a, b, \in G$.

(b) Give an example of a group G, an integer n, and elements $a, b \in G$ such that $(ab)^n \neq a^n b^n$.

5. (a) Let G be a group with the property that $(ab)^2 = a^2 b^2$ for all $a, b \in G$. Prove that G is abelian.

(b) Let G be a group with the property that $(ab)^i = a^i b^i \ \forall a, b \in G$ for three consecutive values of i. Prove that G is abelian.

(c) Show that the conclusion of part (b) does not follow if we only assume that $(ab)^i = a^i b^i$ for two consecutive values of i.

6. Let G be a finite group with an even number of elements. Prove that there is an element $x \in G$ with the property that $x \neq e$ and $x^2 = e$.

7. Let $D = \{(x, \varepsilon) : x \in \{1, -1, i, -i\}, \varepsilon \in \{1, -1\}\}$, and define a binary operation $*$ on D by

$$(x, \varepsilon) * (y, \delta) = (xy^\varepsilon, \varepsilon\delta)$$

for all $x, y \in \{1, -1, i, -i\}$ and $\varepsilon, \delta \in \{1, -1\}$. So for example

$$(-i, -1) * (i, 1) = (-i \cdot i^{-1}, -1 \cdot 1) = (-1, -1).$$

(i) Prove that $(D, *)$ is a group of size 8, with identity element $e = (1, 1)$.

(ii) Let $a = (i, 1)$ and $b = (1, -1)$. Show that $a^4 = b^2 = e$, and that

$$D = \{e, a, a^2, a^3, b, ab, a^2 b, a^3 b\}.$$

(iii) Which element in the list in (ii) is equal to ba? Is $(D, *)$ abelian?

(iv) Work out the group table of $(D, *)$.

(v) For each of the integers $n = 0, 1, 2, 3, 4$, calculate the number of elements $x \in D$ such that $x^n = e$.

8. Still a film critic by night, Ivor Smallbrain has taken up a day job as Head of Binary Operations for the huge poetry production company *Identity in Verse*. At the end of one sunny Tuesday, he notices that the number of aspiring poets who have been in his office is equal to the number of different binary operations on the set $S = \{a, b\}$, while the number who were actually hired is equal to the number of these binary operations that make S into a group.

How many aspiring poets were there on that sunny Tuesday, and how many were hired?

Calculate these numbers for the following wet Friday, when Ivor found them to be defined in the same way, but replacing S by the set $T = \{x, y, z\}$.

Chapter 26

Introduction to Abstract Algebra: More on Groups

So, with some sadness, we have arrived at the final chapter. Fear not, it is also one of the most exciting. In it, you will begin to see just how far one can get using just the four group axioms. For example, we'll prove that if you have a finite group G, and you take any of its elements and raise it to the $|G|^{th}$ power, you always get the identity. We shall also see how groups can be applied in number theory — in particular, to help in finding the largest prime numbers known to mankind.

Subgroups

One of the keys to studying groups is to investigate their "subgroups":

DEFINITION *Let $(G, *)$ be a group and let H be a subset of G. We say that H is a* subgroup *of $(G, *)$ if H is itself a group under $*$.*

Example 26.1

1. $(\mathbb{Z}, +)$ is a subgroup of $(\mathbb{R}, +)$.

2. $(\mathbb{Q}^*, \times)$ is *not* a subgroup of $(\mathbb{R}, +)$, since according to the definition, the binary operation must be the same for the subgroup as for the group in which it lies.

3. $(\{1, -1\}, \times)$ is a subgroup of $(\{1, -1, i, -i\}, \times)$.

4. $(\{1, i\}, \times)$ is *not* a subgroup of $(\{1, -1, i, -i\}, \times)$, since the closure axiom fails ($i \times i = -1$. which is not in the subset).

Here is a useful criterion for deciding whether a given subset of a group is a subgroup. We adopt the multiplicative notation introduced at the end of the previous chapter.

PROPOSITION 26.1
Let G be a group, and let H be a subset of G. The H is a subgroup of G if the following three conditions hold:

(1) $e \in H$ *(where e is the identity element of G),*
(2) $x, y \in H \Rightarrow xy \in H$,
(3) $x \in H \Rightarrow x^{-1} \in H$.

PROOF   Assume that (1)-(3) hold. We check the group axioms for H. The closure axiom holds by (2). The associativity axiom holds for H, since it holds for G. The element e is in H by (1), and is an identity element for H. Finally, the inverse axiom holds for H by (3).   ∎

Example 26.2
(1) Any group G is a subgroup of itself. So is the subset $\{e\}$ consisting of just the identity element.

(2) Let G be the group $(\mathbb{C}^*, \times)$ of nonzero complex numbers under multiplication, and let n be a positive integer. Define

$$C_n = \{z \in \mathbb{C} : z^n = 1\},$$

the set of nth roots of 1 in $\mathbb{C}$. Certainly C_n is a subset of G. Let's check conditions (1)-(3) of the proposition. Certainly $1 \in C_n$ so (1) holds. If $y, z \in C_n$ then $(yz)^n = y^n z^n = 1$, so $yz \in C_n$; and if $z \in C_n$ then $(z^{-1})^n = (\frac{1}{z})^n = \frac{1}{z^n} = 1$, so $z^{-1} \in C_n$. Thus (2) and (3) hold also, and so C_n is a subgroup of G. Note that by Proposition 6.3, if $\omega = e^{2\pi i/n}$ then

$$C_n = \{1, \omega, \omega^2, \ldots, \omega^{n-1}\}$$

so C_n has size n.
   Notice that C_2 is the subgroup $\{1, -1\}$, and $C_4 = \{1, -1, i, -i\}$, as in Examples 25.2(5,6).

(3) Let n be a positive integer. Recall from Chapter 20 the definition of an *even* permutation in the symmetric group S_n, and define

$$A_n = \{g \in S_n : g \text{ even}\} = \{g \in S_n : \text{sgn}(g) = 1\}.$$

We use Proposition 20.5 to check the subgroup conditions for A_n. The identity element e is even, by Proposition 20.5(i); if $g, h \in A_n$ then by

Proposition 20.5(ii), $\text{sgn}(gh) = \text{sgn}(g)\,\text{sgn}(h) = 1$, so $gh \in A_n$; and if $g \in A_n$ then by Proposition 20.5(iii), $\text{sgn}(g^{-1}) = \text{sgn}(g) = 1$, so $g^{-1} \in A_n$. Hence A_n is a subgroup of S_n.

We call the subgroup A_n in Example 26.2(3) the *alternating group of degree n*. By Exercise 5 in Chapter 20, exactly half the permutations in S_n are even, and hence

$$|A_n| = \frac{1}{2}n!.$$

For example $|A_3| = 3$, and A_3 consists of the permutations e, $(1\,2\,3)$ and $(1\,3\,2)$. More interesting is the group A_4: it has size 12, and consists of all the permutations in S_4 of cycle-shapes $(3,1)$ and (2^2) together with the identity.

The next result provides a large supply of subgroups of any group.

PROPOSITION 26.2
Let G be a group, and let $a \in G$. Define

$$A = \{a^n \,|\, n \in \mathbb{Z}\} = \{\ldots, a^{-2}, a^{-1}, a^0, a, a^2, \ldots\}.$$

Then A is a subgroup of G.

PROOF   We check the conditions (1)-(3) in Proposition 26.1. For (1), note that $e = a^0 \in A$. For (2), let $a^m, a^n \in A$; then by Proposition 25.3, $a^m a^n = a^{m+n} \in A$. Finally, if $a^n \in A$ then $(a^n)^{-1} = a^{-n} \in A$, proving (3).   ∎

DEFINITION   *In the notation of the previous proposition, we write $A = \langle a \rangle$, and call it the* cyclic subgroup of G generated by a. *So for each element $a \in G$ there is a cyclic subgroup $\langle a \rangle$ of G.*

Example 26.3
(1) In the group $(\mathbb{Z}, +)$, what is the cyclic subgroup $\langle 3 \rangle$? Well, in this group under the multiplicative notation we have $3^1 = 3$, $3^2 = 3 + 3 = 6$, and $3^n = 3 + \cdots + 3 = 3n$ for $n > 0$. Similarly $3^{-n} = -3n$. Hence

$$\langle 3 \rangle = \{3n : n \in \mathbb{Z}\},$$

the set of multiples of 3. There is nothing special about the number 3 in this example: for any integer a, the cyclic subgroup $\langle a \rangle = \{an : n \in \mathbb{Z}\}$. In particular, $\langle 1 \rangle = \mathbb{Z}$.

(2) Let $G = (\mathbb{C}^*, \times)$ and let $\omega = e^{2\pi i/n}$, where n is a positive integer. The cyclic subgroup of G generated by ω consists of all the powers ω^k

($k \in \mathbb{Z}$). Each of these is an nth root of 1, so is equal to one of the powers $1, \omega, \ldots, \omega^{n-1}$ by Proposition 6.3. Hence

$$\langle \omega \rangle = \{1, \omega, \ldots, \omega^{n-1}\}.$$

This is the group C_n of Example 26.2(2).

(3) Let $G = S_3$, the symmetric group of degree 3, and let $a = (123) \in G$. What is $\langle a \rangle$? Well, it contains $a^0 = e$, $a^1 = a$ and $a^2 = (132)$. We know that $a^3 = e$, so $a^4 = a, a^5 = a^2$, etc. and we don't get any further elements by taking positive powers of a. Also $a^{-1} = a^3 a^{-1} = a^2$, so $a^{-2} = a$, $a^{-3} = e$ and so on. We conclude that

$$\langle a \rangle = \{e, a, a^2\}.$$

Now consider the cyclic subgroup $\langle b \rangle$, where $b = (12)$. Here $b^0 = e$, $b^1 = b^{-1} = b$, $b^2 = b^{-2} = e$ and so on, so

$$\langle b \rangle = \{e, b\}.$$

(4) Recall from Example 25.3 that $S_3 = \{e, a, a^2, b, ab, a^2b\}$. Here is a list of all the cyclic subgroups of S_3:

$$\langle e \rangle = \{e\}$$
$$\langle a \rangle = \{e, a, a^2\}$$
$$\langle a^2 \rangle = \{e, a, a^2\}$$
$$\langle b \rangle = \{e, b\}$$
$$\langle ab \rangle = \{e, ab\}$$
$$\langle a^2b \rangle = \{e, a^2b\}.$$

(5) Let $G = S_4$, and let V be the following subset of G:

$$V = \{e, (12)(34), (13)(24), (14)(23)\}.$$

You are asked to show that V is a subgroup of G in Exercise 2 at the end of the chapter. It is *not* a cyclic subgroup, since as in Example (3) above we can see that the cyclic subgroup generated by each of its non-identity elements has size 2, whereas $|V| = 4$.

DEFINITION   We say that a group G is a *cyclic group* if there exists an element $a \in G$ such that $G = \langle a \rangle$. If this is the case we call a a *generator* for G.

For example, if $G = (\mathbb{Z}, +)$ then from Example 26.3(1) we have $G = \langle 1 \rangle$; so $(\mathbb{Z}, +)$ is cyclic with generator 1. The group C_n of nth roots of 1 in $\mathbb{C}^*$ is cyclic with generator $\omega = e^{2\pi i/n}$ (see Example 26.3(2)). On the other hand, S_3 is *not* cyclic, as it is not equal to any of its cyclic subgroups (see Example 26.3(4)).

Order of an Element

DEFINITION   *Let G be a group, and let $a \in G$. The order of a, written $o(a)$, is the smallest positive integer k such that $a^k = e$. If no such k exists, we say that a has infinite order and write $o(a) = \infty$.*

Thus saying that $o(a) = k$ with k finite means that $a^k = e$ and $a^i \neq e$ for $1 \leq i \leq k - 1$. On the other hand, saying that $o(a) = \infty$ means that $a^i \neq e$ for all $i > 0$.

Example 26.4
(1) The identity element e has order 1, and is the only such element.

(2) We already introduced the order of a permutation $g \in S_n$ in Chapter 20, and showed in Proposition 20.4 that when g is expressed as a product of disjoint cycles, $o(g)$ is equal to the least common multiple of the lengths of the cycles. For example for $a, b \in S_3$ as in Example 26.3(3) above, we have

$$o(a) = o(a^2) = 3, \ o(b) = o(ab) = o(a^2 b) = 2.$$

(3) For $G = (\mathbb{Z}, +)$ what is $o(3)$, the order of the element $3 \in G$? Well, in G the identity element is 0, and 3^n is equal to the integer $3n$. So $3^n \neq e$ for any positive integer n, and hence $o(3) = \infty$.

PROPOSITION 26.3
Let G be a group and let $a \in G$. The number of elements in the cyclic subgroup $\langle a \rangle$ is equal to $o(a)$.

PROOF   (A) First assume that $o(a) = k$ is finite. So $a^k = e$ but $a^i \neq e$ for $1 \leq i \leq k - 1$. Let $A = \langle a \rangle = \{a^n : n \in \mathbb{Z}\}$. Certainly A contains the elements

$$e, a, a^2, \ldots, a^{k-1}. \tag{26.1}$$

We claim that these elements are all different. To see this, suppose it is false, so there exist i, j with $1 \leq i < j \leq k - 1$ such that $a^i = a^j$. But then

$$a^i = a^j \Rightarrow a^{-i} a^i = a^{-i} a^j \Rightarrow e = a^{j-i},$$

and this is a contradiction since $1 \leq j - i \leq k - 1$. Therefore, as claimed, the list in (26.1) consists of k different elements in A. So $|A| \geq k$.

We now show that every element of A is in the list (26.1). Let $a^n \in A$ with $n \in \mathbb{Z}$. By Proposition 10.1 there are integers q, r such that $n = qk + r$ and $0 \leq r < k$. Then

$$
\begin{aligned}
a^n &= a^{qk+r} \\
&= a^{qk}a^r \\
&= (a^k)^q a^r \\
&= e^q a^r \\
&= a^r.
\end{aligned}
$$

Hence a^n is in the list (26.1). We conclude that $A = \langle a \rangle = \{e, a, a^2, \ldots, a^{k-1}\}$ and $|A| = k = o(a)$, proving the proposition in the case where $o(a)$ is finite.

(B) Now assume that $o(a) = \infty$. This means that $a^i \neq e$ for all $i > 0$. If $a^i = a^j$ with $i < j$, then as above we see that $e = a^{j-i}$, which is a contradiction. Hence $A = \langle a \rangle = \{\ldots, a^{-2}, a^{-1}, e, a, a^2, \ldots\}$ and all these elements are distinct. It follows that $|A| = \infty = o(a)$, completing the proof. ∎

Example 26.5

(1) Let $G = S_3$ and $a = (1\,2\,3), , b = (1\,2) \in G$ as in Example 26.3(3). Then $o(a) = 3$, $\langle a \rangle = \{e, a, a^2\}$ and $o(b) = 2$, $\langle b \rangle = \{e, b\}$.

(2) In $(\mathbb{Z}, +)$ we observed that $o(3) = \infty$ and $\langle 3 \rangle = \{3n : n \in \mathbb{Z}\}$, an infinite group.

(3) For $\omega = e^{2\pi i/n} \in \mathbb{C}^*$, the cyclic group $C_n = \langle \omega \rangle = \{1, \omega, \ldots, \omega^{n-1}\}$ has size n, and $o(\omega) = n$.

PROPOSITION 26.4
Let G be a group and let $a \in G$. Suppose $a^n = e$, where n is a positive integer. Then $o(a)$ divides n.

PROOF   Let $k = o(a)$. There are integers q, r such that $n = qk + r$ and $0 \leq r < k$. Now

$$
e = a^n = a^{qk+r} = (a^k)^q a^r = e^q a^r = a^r.
$$

So $a^r = e$. Since k is the smallest positive integer such that $a^k = e$, and $r < k$, it follows that $r = 0$. Therefore $n = qk$, showing that k divides n. ∎

Lagrange's Theorem

This is a famous result which relates the number of elements in a subgroup of a finite group G to the number of elements in G. Recall that $|G|$ denotes the size of G — that is, the number of elements in G.

THEOREM 26.1 (Lagrange's Theorem)

Let G be a finite group, and let H be a subgroup of G. Then $|H|$ divides $|G|$.

For example, the theorem tells us that any subgroup of S_3 must have size 1,2,3 or 6, and any subgroup of C_5 has size 1 or 5.

Note that Lagrange's theorem is a *one-way implication*:

$$H \text{ a subgroup of } G \Rightarrow |H| \text{ divides } |G|.$$

It does not hold the other way round, since there exists a group G and a positive integer r dividing G such that G has no subgroup of size r. The smallest instance of this is for the alternating group A_4; this has size 12, but it can be proved that it has no subgroup of size 6 (see Exercise 8 at the end of the chapter).

It may not be immediately apparent, but Lagrange's theorem has some strong consequences, both in group theory and some of its applications. I will postpone proving the theorem and first show you some of these consequences.

Consequences of Lagrange's Theorem

Here are three group-theoretic consequences.

COROLLARY 26.1

Let G be a finite group and let $a \in G$. Then $o(a)$ is finite and divides $|G|$.

PROOF   Let $A = \langle a \rangle$, the cyclic subgroup of G generated by a. By Proposition 26.3, $|A| = o(a)$. Hence $o(a)$ is finite, and it divides $|G|$ by Lagrange's Theorem. ∎

COROLLARY 26.2
Let G be a finite group and let $N = |G|$. Then $a^N = e$ for all $a \in G$.

PROOF   Let $k = o(a)$. By Corollary 26.1, k is finite and divides N. So $N = kr$ for some integer r. Then

$$a^N = a^{kr} = (a^k)^r = e^r = e.$$

∎

I find this result really quite remarkable — if you have a finite group G, and you take any of its elements and combine it with itself $|G|$ times, you always get the identity. Think about it — would you have guessed such a thing to be true, just looking at the group axioms?

COROLLARY 26.3
Let G be a group, and suppose that $|G|$ is a prime number. Then G is a cyclic group.

PROOF   Let $|G| = p$, a prime. Since $p \geq 2$, there exists $a \in G$ such that $a \neq e$. By Lagrange's Theorem, the cyclic subgroup $A = \langle a \rangle$ has size dividing p. Since A contains e and a we have $|A| \geq 2$, and hence $|A| = p$ as p is prime. As $|A| = |G|$ it follows that $G = A = \langle a \rangle$, so G is cyclic.

∎

Example 26.6
By Lagrange's Theorem any subgroup of S_3 must have size 1,2,3 or 6. There are unique subgroups of sizes 1 and 6, namely $\{e\}$ and S_3 itself. Any other subgroup has size 2 or 3. Since 2 and 3 are primes, such subgroups are cyclic, by Corollary 26.3. Hence they are all in the list in Example 26.3(4). So we know *all* the subgroups of S_3.

Applications to Number Theory

In this section I will show you some applications of group theory to number theory (i.e., the theory of the integers, primes, etc.). The key to this is the system $\mathbb{Z}_m$ defined at the end of Chapter 13. Assume that $m \geq 2$. Recall that $\mathbb{Z}_m = \{\bar{0}, \bar{1}, \bar{2}, \ldots, \overline{m-1}\}$, and we defined addition and multiplication on $\mathbb{Z}_m$ as follows: for $\bar{x}, \bar{y} \in \mathbb{Z}_m$, the sum $\bar{x} + \bar{y}$ and product $\bar{x}\bar{y}$ are the elements $\bar{k}, \bar{l}$ of $\mathbb{Z}_m$

such that k, l are congruent to the integers $x + y$ and xy modulo m, respectively. So we have two natural binary operations $+$ and $\times$ defined on $\mathbb{Z}_m$. Can we get groups out of these?

It is rather easy to see that $(\mathbb{Z}_m, +)$ is a group. The closure axiom holds, the identity is $\bar{0}$, the inverse of $\bar{x}$ is $\overline{m - x}$. Finally, $(\bar{x} + \bar{y}) + \bar{z} = \bar{r}$, where $(x + y) + z \equiv r \bmod m$; since $(x + y) + z = x + (y + z)$ (these are just integers), $\bar{x} + (\bar{y} + \bar{z}) = \bar{r}$ as well, and so the associativity axiom holds.

In fact $(\mathbb{Z}_m, +)$ is not a particularly thrilling group. It is cyclic, with generator $\bar{1}$. Much more interesting is $(\mathbb{Z}_m, \times)$. This satisfies closure, we can show it is associative just as we did for $(\mathbb{Z}_m, +)$, and it has identity element $\bar{1}$. But the inverse axiom fails, since $\bar{0}$ has no inverse (as $\bar{0} \times \bar{x} = \bar{0}$ for all $\bar{x} \in \mathbb{Z}_m$). Let's try to remedy this by excluding $\bar{0}$ and defining

$$\mathbb{Z}_m^* = \mathbb{Z}_m - \{\bar{0}\}.$$

Is this now a group? Well no, certainly not for all values of m — for example in $\mathbb{Z}_4$ we have $\bar{2} \times \bar{2} = \bar{0}$ which is not in $\mathbb{Z}_4^*$, so $\mathbb{Z}_4^*$ is not even closed under multiplication. In fact, if m is not prime, then $m = ab$ for some $a, b \in \{1, \ldots, m-1\}$, and so $\bar{a} \times \bar{b} = \bar{0}$, showing that $\mathbb{Z}_m^*$ is not closed.

So we are left to consider $\mathbb{Z}_p^*$, where p is prime. Here at last there is something positive to say:

PROPOSITION 26.5

Let p be a prime. Then $(\mathbb{Z}_p^*, \times)$ is a group. It is abelian, and has size $p - 1$.

PROOF First we need to check closure. Let $\bar{x}, \bar{y} \in \mathbb{Z}_p^*$. Then $x, y \in \{1, \ldots, p-1\}$, so both are coprime to p. Hence xy is also coprime to p by Proposition 10.5(b), which implies that $\bar{x}\bar{y} \neq \bar{0}$. So $\bar{x}\bar{y} \in \mathbb{Z}_p^*$, proving closure.

We have already observed that $(\mathbb{Z}_p^*, \times)$ is associative and has identity element $\bar{1}$. So it remains to check the inverse axiom. Let $\bar{x} \in \mathbb{Z}_p^*$. Then $\mathrm{hcf}(x, p) = 1$, so by Proposition 13.6, there is an integer y such that $xy \equiv 1 \bmod p$. Certainly $y \not\equiv 0 \bmod p$, and replacing y by its remainder on division by p, we can take it that $y \in \{1, \ldots, p-1\}$. Then $\bar{y} \in \mathbb{Z}_p^*$ and $\bar{x}\bar{y} = \bar{y}\bar{x} = \bar{1}$. Thus $\bar{y}$ is the inverse of $\bar{x}$, proving the inverse axiom.

Thus $(\mathbb{Z}_p^*, \times)$ is a group. From the definition of multiplication, $\bar{x}\bar{y} = \bar{y}\bar{x}$, so this group is abelian, and it has $p - 1$ elements. ∎

Example 26.7

(1) Is $\mathbb{Z}_5^* = \{\bar{1}, \bar{2}, \bar{3}, \bar{4}\}$ a cyclic group? The answer is yes: since $\bar{2}^2 = \bar{4}$, $\bar{2}^3 = \bar{3}$ and $\bar{2}^4 = \bar{1}$, we have $\mathbb{Z}_5^* = \langle \bar{2} \rangle$.

The group $\mathbb{Z}_7^*$ is also cyclic: this time $\bar{2}$ is not a generator (since $\bar{2}^3 = \bar{1}$, so the element $\bar{2}$ has order 3), but you can check that $\bar{3}$ is a generator.

In fact $\mathbb{Z}_p^*$ is always cyclic — I won't prove this elegant and important fact, but refer you to the book by I.N. Herstein listed in the Further Reading at the end of this book.

(2) In the group $\mathbb{Z}_{31}^*$, what is $\overline{11}^{-1}$? To answer this we need to find an integer y such that $11y \equiv 1 \bmod 31$. Now $\text{hcf}(11, 31) = 1$, so by Proposition 10.3 there are integers s, t such that $11s + 31t = 1$. Using the Euclidean algorithm as in Example 10.2, we find that $11 \cdot (-14) + 31 \cdot 5 = 1$. Hence $11 \cdot (-14) \equiv 1 \bmod 31$, and so also $11 \cdot 17 \equiv 1 \bmod 31$. So $\overline{11}^{-1} = \overline{17}$.

Our first application of group theory is another proof of Fermat's Little Theorem (the third one in this little book!).

THEOREM 26.2
Let p be a prime and let a be an integer that is not divisible by p. Then $a^{p-1} \equiv 1 \bmod p$.

PROOF   We shall work in the group $(\mathbb{Z}_p^*, \times)$. Let $r \in \{1, \ldots p-1\}$ be such that $a \equiv r \bmod p$. Then $\bar{r} \in \mathbb{Z}_p^*$. Now $\mathbb{Z}_p^*$ has size $p-1$, so by Corollary 26.2, $\bar{r}^{p-1} = \bar{1}$. This means that $r^{p-1} \equiv 1 \bmod p$, and hence (by Proposition 13.4) that $a^{p-1} \equiv 1 \bmod p$. ∎

It is possible to generalize the construction of $\mathbb{Z}_p^*$, as follows. Let m be an arbitrary positive integer greater than 1, and for each x such that $1 \le x \le m-1$ and $\text{hcf}(x, m) = 1$, define a symbol $\bar{x}$. Let

$$U(\mathbb{Z}_m) = \{\bar{x} : 1 \le x \le m-1 \text{ and } \text{hcf}(x, m) = 1\}.$$

Define multiplication on $U(\mathbb{Z}_m)$ by $\bar{x}\bar{y} = \bar{k}$, where $xy \equiv k \bmod m$ and $1 \le k \le m-1$.

PROPOSITION 26.6
$(U(\mathbb{Z}_m), \times)$ *is a group. It is abelian, and has size $\phi(m)$, where ϕ is Euler's ϕ-function.*

PROOF   The proof that $U(\mathbb{Z}_m)$ is a group is almost exactly the same as that of Proposition 26.5; I have set it as Exercise 9 at the end of the chapter to make sure you fill in the details. It is clearly abelian, and has size $\phi(m)$ by the definition of the Euler ϕ-function (see Chapter 17). ∎

You may find the notation $U(\mathbb{Z}_m)$ a bit strange — what is the letter U doing there? In fact it is quite a natural notation, as this is the group of "units" in $\mathbb{Z}_m$ (where a unit is defined to be an element that has an inverse).

Example 26.8

(1) For p prime, $U(\mathbb{Z}_p)$ is just the group $\mathbb{Z}_p^*$.

(2) $U(\mathbb{Z}_4) = \{\bar{1}, \bar{3}\}$, $U(\mathbb{Z}_6) = \{\bar{1}, \bar{5}\}$ and $U(\mathbb{Z}_8) = \{\bar{1}, \bar{3}, \bar{5}, \bar{7}\}$. Notice that $U(\mathbb{Z}_8)$ is not cyclic, as all its non-identity element have order 2.

Copying the proof of Theorem 26.2, replacing the group $\mathbb{Z}_p^*$ by $U(\mathbb{Z}_m)$, gives the following result.

PROPOSITION 26.7

Let m be a positive integer, and let a be an integer which is coprime to m. Then
$$a^{\phi(m)} \equiv 1 \ mod \ m.$$

For example this shows that $a^4 \equiv 1$ mod 8 for all odd integers a, and so on. Notice also that Proposition 14.1 is a special case of the above result, since $\phi(pq) = (p-1)(q-1)$ when p, q are distinct primes.

Before moving on, let me point out another pleasing application of the group $\mathbb{Z}_p^*$. Recall from Proposition 3.4 that the decimal expansion of any rational number $\frac{1}{n}$ is periodic. Can we say anything about the relationship between the period and the number n? For example, here are the periods of the decimal expansions of the rationals $\frac{1}{p}$ for the first few primes p:

p	2	3	5	7	11	13	17
period of $\frac{1}{p}$	1	1	1	6	2	6	16

What is going on? Fermat himself considered this question, and came up with the following result.

PROPOSITION 26.8

If p is prime, then the period of the decimal expansion of the rational $\frac{1}{p}$ divides $p - 1$.

PROOF   This is true for $p = 2$ or 5, so assume $p \notin \{2, 5\}$. Then p is coprime to 10. To express $\frac{1}{p}$ as a decimal we perform long division of p into $1.000\ldots$. At each stage of the division we get a remainder x, which we regard as the element $\bar{x}$ of the group $\mathbb{Z}_p^*$. So the first remainder is $\bar{1} \in \mathbb{Z}_p^*$, the next is $\overline{10}$, the next is $\overline{10}^2$, and so on. The sequence of decimal digits will start repeating the first time this sequence of group elements $\bar{1}, \overline{10}, \overline{10}^2, \ldots$ arrives back at $\bar{1}$. So the period is equal to the smallest postive integer k such that $\overline{10}^k = \bar{1}$ in the group $\mathbb{Z}_p^*$. In other words, it is the order of the element $\overline{10}$ in $\mathbb{Z}_p^*$. By Corollary 26.1, this order divides $p-1$. ∎

Naturally, Fermat's proof was somewhat different to this one, as group theory was not around in his time.

A very similar proof shows that for any positive integer n that is coprime to 10, the period of $\frac{1}{n}$ divides $\phi(n)$ (Exercise 14 at the end of the chapter).

Our next application of group theory concerns some special kinds of prime numbers.

DEFINITION   *A prime number p is called a* Mersenne prime *if $p = 2^n - 1$ for some positive integer n.*

The first few Mersenne primes are

$$2^2 - 2 = 3,$$
$$2^3 - 1 = 7,$$
$$2^5 - 1 = 31,$$
$$2^7 - 1 = 127.$$

Notice that $2^4 - 1$ and $2^6 - 1$ are not prime. This is readily explained by the next result.

PROPOSITION 26.9
Suppose $2^n - 1$ is prime. Then n is prime.

PROOF   Suppose n is not prime. Then $n = ab$ for some integers $a, b > 1$. Hence $2^n - 1 = 2^{ab} - 1$ has a factor $2^a - 1$ (just take $x = 2^a$ in the equation $x^b - 1 = (x-1)(x^{b-1} + x^{b-2} + \cdots + 1)$). Since $1 < 2^a - 1 < 2^n - 1$, this means that $2^n - 1$ is not prime, which is a contradiction. Therefore n is prime. ∎

Thus Mersenne primes are necessarily of the form $2^p - 1$, where p is itself prime. Interest in Mersenne primes stems from the fact that there are much more powerful methods for testing whether a number of the form $2^p - 1$ is prime than just the general primality testing methods discussed at the end of Chapter 14. Indeed, at the time I am writing this (April 2015), the largest explicitly known prime is $2^{57,885,161} - 1$, a number with $17,425,170$ digits.

Possibly at the time you are reading this a larger prime will have been found, but it is almost certain to be a Mersenne prime.

I will show you one of the methods for testing whether $2^p - 1$ is prime that comes from group theory. Before doing that, let me point out a connection with another type of number.

DEFINITION *A positive integer N is perfect if N is equal to the sum of its positive divisors (including 1 but not N).*

The smallest two perfect numbers are

$$6 = 1 + 2 + 3,$$
$$28 = 1 + 2 + 4 + 7 + 14.$$

A very neat fact, going back to the ancient Greeks, is that if $2^p - 1$ is prime then the number $2^{p-1}(2^p - 1)$ is perfect. I have set this as Exercise 12 at the end of the chapter. For example, $6 = 2(2^2 - 1)$ and $28 = 2^2(2^3 - 1)$ are of this form; the next is $2^4(2^5 - 1) = 496$. Far less obvious is the following partial converse, proved by Euler in 1849: every *even* perfect number is of the form $2^{p-1}(2^p - 1)$, where $2^p - 1$ is prime (for a proof of this, see the book by K.H. Rosen listed in the Further Reading at the end of this book). It is unknown to this day whether there exist any *odd* perfect numbers.

Our group-theoretic method for testing for Mersenne primes is based on the next result.

PROPOSITION 26.10
Let p be a prime, and let $N = 2^p - 1$. Suppose q is a prime divisor of N. Then $q \equiv 1 \bmod p$.

PROOF We shall work in the group $(\mathbb{Z}_q^*, \times)$. Since q divides $2^p - 1$ we know that $2^p \equiv 1 \bmod q$. This means that in $\mathbb{Z}_q^*$ we have $\bar{2}^p = \bar{1}$. Hence by Proposition 26.4, the order $o(\bar{2})$ of the element $\bar{2}$ divides p. As p is prime, $o(\bar{2})$ is therefore either 1 or p. If it is 1, then $\bar{2} = \bar{1}$ in $\mathbb{Z}_q^*$, which is a contradiction. Hence $o(\bar{2}) = p$. By Corollary 26.1, this implies that p divides the size of the group $\mathbb{Z}_q^*$, which is equal to $q - 1$. In other words, $q \equiv 1 \bmod p$. ∎

Example 26.9
(1) Is $2^{11} - 1$ prime? To answer this, let $N = 2^{11} - 1 = 2047$. If N is not prime, then it is divisible by a prime $q \le \sqrt{N}$ (see Exercise 2 of Chapter 12), and $q \equiv 1 \bmod 11$ by Proposition 26.10. The numbers less than $\sqrt{N}$

that are congruent to 1 modulo 11 are

$$12, 23, 34, 45$$

and of these, only 23 is prime. So the only possibility for q is 23. It remains to check whether or not $2^{11} \equiv 1 \bmod 23$. In fact this is true! (Check this using the method of Example 13.3.) Hence 23 divides N and so N is not prime.

(2) The next two possibilities are $2^{13} - 1$ and $2^{17} - 1$. Both are prime — I have set this in Exercise 12 at the end of the chapter.

Based on the evidence so far, one might think that Mersenne primes are quite common. But that appears not to be the case — for example, of the 168 primes p less than 1000, the number $2^p - 1$ is prime for only 14 of them. It is another unsolved problem in prime number theory (to go with those described at the end of Chapter 12) whether or not there are infinitely many Mersenne primes.

Proof of Lagrange's Theorem

We've deduced all sorts of things using Lagrange's Theorem. I will conclude the chapter by proving it.

Let G be a finite group, and let H be a subgroup of G. Let $m = |H|$ and $H = \{h_1, \ldots, h_m\}$.

First let me give you the idea behind the proof. Write down the m elements of H in a column. Now choose an element $x \in G - H$, and write down the elements $h_1x, h_2x, \ldots h_mx$ in another column. Next choose an element y not listed so far (if one exists), and write down the elements $h_1y, h_2y, \ldots h_my$ in a third column. Carry on doing this until we run out of elements of G (which will happen, as G is finite). This process lists the elements of G in an array as follows:

H	Hx	Hy	$\cdots$
h_1	h_1x	h_1y	$\cdots$
h_2	h_2x	h_2y	$\cdots$
$\vdots$	$\vdots$	$\vdots$	
h_m	h_mx	h_my	$\cdots$

The aim is to show that the columns of this array have the following properties:

(A) each column has m distinct elements

(B) the columns form a partition of G — that is, every element of G belongs to exactly one columnn.

Given (A) and (B), it follows that the total number of elements of G is equal to rm, where r is the number of columns. Hence m divides $|G|$, which is the conclusion of Lagrange's Theorem.

So that's the idea. Let's now carry it out. The columns in the above array are very important subsets of G, and we give them a special name:

DEFINITION   *For $x \in G$, define*

$$Hx = \{hx : h \in G\} = \{h_1x, \ldots, h_mx\}.$$

This is a subset of G, called a right *coset of H in G.*

Example 26.10
(1) Since $He = H$, the subgroup H itself is a right coset.

(2) Let $G = S_3$, and let $a = (1\,2\,3)$, $b = (1\,2)$. Define H to be the subgroup $\langle a \rangle = \{e, a, a^2\}$ of G. Let's write down all the right cosets of H in G. Recall that $G = \{e, a, a^2, b, ab, a^2b\}$. Check that

$$He = Ha = Ha^2 = \{e, a, a^2\},$$
$$Hb = Hab = Ha^2b = \{b, ab, a^2b\}.$$

So there are just 2 different right cosets. Both have size 3, and they form a partition of G.

(3) Now let K be the subgroup $\langle b \rangle = \{e, b\}$ of $G = S_3$. The right cosets of K in G are

$$Ke = Kb = \{e, b\},$$
$$Ka = Ka^2b = \{a, a^2b\},$$
$$Ka^2 = Kab = \{a^2, ab\}.$$

There are 3 different right cosets, all of size 2, and they also form a partition of G.

Now let's start the proof proper of Lagrange's Theorem. As above, let G be a finite group and let $H = \{h_1, \ldots, h_m\}$ be a subgroup of size m. The next three results are aimed at proving properties (A) and (B) described above for the collection of right cosets of H in G.

PROPOSITION 26.11
For any $x \in G$ we have $|Hx| = m$; that is, the right coset Hx has m elements.

PROOF By definition, $Hx = \{h_1 x, \ldots, h_m x\}$. If $h_i x = h_j x$, then $h_i = h_j$ by Proposition 25.2. Hence the elements $h_1 x, \ldots, h_m x$ are all distinct, and so $|Hx| = m$. ∎

PROPOSITION 26.12
Let $x, y \in G$. Then either $Hx = Hy$ or $Hx \cap Hy = \emptyset$.

PROOF Assume that $Hx \cap Hy \neq \emptyset$. We show that this implies that $Hx = Hy$, which will prove the proposition.

By the assumption there is an element $a \in Hx \cap Hy$. So there exist $h_i, h_j \in H$ such that $a = h_i x = h_j y$. Then

$$h_i x = h_j y \Rightarrow h_i^{-1} h_i x = h_i^{-1} h_j y \Rightarrow x = h_i^{-1} h_j y.$$

Hence for any $h \in H$ we have $hx = hh_i^{-1} h_j y$. As H is a subgroup, closure in H shows that $hh_i^{-1} h_j \in H$, so it follows that $hx \in Hy$. So we have shown that $Hx \subseteq Hy$.

Similarly, $y = h_j^{-1} h_i x$, so for $h \in H$ we have $hy = hh_j^{-1} h_i x \in Hx$. This shows that $Hy \subseteq Hx$. We now have $Hx \subseteq Hy$ and $Hy \subseteq Hx$. Hence $Hx = Hy$, proving the result. ∎

PROPOSITION 26.13
Let $x \in G$. Then x lies in the right coset Hx.

PROOF As H is a subgroup, $e \in H$. So $x = ex \in Hx$. ∎

Completion of the proof

We are now ready to complete the proof of Lagrange's Theorem. By Proposition 26.13, G is equal to the union of all the right cosets of H — that is,

$$G = \bigcup_{x \in G} Hx.$$

Some of these right cosets will be equal (e.g., in Example 26.10(2) above, $He = Ha = Ha^2$). Let the list of *different* right cosets be $Hx_1, \ldots, Hx_r$. Then

$$G = Hx_1 \cup Hx_2 \cup \cdots \cup Hx_r,$$

and $Hx_i \neq Hx_j$ if $i \neq j$. By Proposition 26.12, $Hx_i \cap Hx_j = \emptyset$ if $i \neq j$. So the right cosets $Hx_1, \ldots, Hx_r$ form a partition of G. Hence

$$|G| = |Hx_1| + \cdots + |Hx_r|.$$

Also $|Hx_i| = m$ for all i, by Proposition 26.11. It follows that $|G| = rm = r|H|$. Therefore $|H|$ divides $|G|$, completing the proof of Lagrange's Theorem.

Remark The above proof can be recast in terms of a neat equivalence relation $\sim$ on G, defined as follows: for $x, y \in G$ define $x \sim y$ if and only if $xy^{-1} \in H$. Showing that $\sim$ is an equivalence relation is Exercise 14(a) at the end of the chapter. The equivalence class of an element $x \in G$ is the right coset Hx (Exercise 14(b)). Hence by Proposition 18.1, the distinct right cosets form a partition of G. Now we can complete the proof using Proposition 26.11 as above.

I hope you enjoyed this little foray into the world of abstract algebra. If it has caught your fancy, have a look at the book by I.N. Herstein listed in the Further Reading at the end of this book.

It is hard to think of a better place to stop, so I will stop.

Exercises for Chapter 26

1. Which of the following subsets H are subgroups of the given group G?

   (i) $G = (\mathbb{Z}, +)$, $H = \{n \in \mathbb{Z} : n \equiv 0 \bmod 37\}$.

   (ii) $G = S_4$, $H = \{x \in G : x^2 = e\}$.

   (iii) $G = (\mathbb{C}, +)$, $H = \{ri : r \in \mathbb{R}\}$.

   (iv) $G = (\mathbb{R}^*, \times)$, $H = \{\pi^n : n \in \mathbb{Z}\}$.

   (v) $G = S_n$, $H = \{g \in G : g(1) = 1\}$ (the set of all permutations that send $1 \rightarrow 1$).

   (vi) $G = S_n$, $H = \{g \in G : g(1) = 2\}$.

   (vii) $G = S_n$, $H = \{g \in G : g(i) - g(j) \equiv i - j \bmod n, \forall i, j\}$.

2. Let V, W and X be the following subsets of S_4:

   $$V = \{e, (12)(34), (13)(24), (14)(23)\}$$
   $$W = \{e, (12), (34), (12)(34)\}$$
   $$X = \{e, (1234), (1432), (13)(24)\}.$$

   Prove that V, W and X are all subgroups of S_4 and decide which of them are cyclic.

3. Prove that every cyclic group is abelian.

   Show that the converse is false by giving an example of an abelian group that is not cyclic.

4. Let G be a finite group, and r a positive integer coprime to $|G|$. Prove that if $x \in G$ satisfies $x^r = e$, then $x = e$.

5. Let G be a group with subgroups H and K.

   (a) Prove that $H \cap K$ is a subgroup of G.

   (b) Show that $H \cup K$ is not a subgroup unless either $H \subseteq K$ or $K \subseteq H$.

   (c) Find an example of a group G with three subgroups H, K, L, none of them equal to G, such that $G = H \cup K \cup L$.

6. Which of the following groups are cyclic?

   (i) $(\mathbb{Q}, +)$.

   (ii) $(\mathbb{Q}^*, \times)$.

   (iii) the subgroup $\langle 4 \rangle \cap \langle 6 \rangle$ of $(\mathbb{Z}, +)$.

   (iv) the symmetric group S_4.

   (v) the alternating group A_3.

   (vi) $(\{\frac{1}{2^n} : n \in \mathbb{Z}\}, \times)$.

7. (a) Write down all the generators of the following cyclic groups:

$$(\mathbb{Z}, +), \ C_4, C_5, C_6.$$

   (b) Prove that the total number of generators of the cyclic group C_n is equal to $\phi(n)$ (where ϕ is the Euler ϕ-function).

8. Let $G = A_4$, the alternating group of degree 4.

   (a) Write down all the elements of order 2 in G.

   (b) Find subgroups of G of sizes 1, 2, 3, 4 and 12.

   (c) Now suppose H is a subgroup of G and $|H| = 6$.

   (i) Show that H has an element x of order 2.

   (ii) Show that H also has an element y of order 3.

   (iii) By considering products involving x and y, obtain a contradiction.

   (Hence A_4 has no subgroup of size 6, showing that the converse of Lagrange's theorem is not true in general.)

9. Let m be a positive integer, and define $U(\mathbb{Z}_m)$ as in the text before Proposition 26.6.

   (a) Prove that $U(\mathbb{Z}_m)$ is a group under multipliciation.

   (b) Which of the groups $U(\mathbb{Z}_9)$, $U(\mathbb{Z}_{10})$, $U(\mathbb{Z}_{12})$ are cyclic?

   (c) Prove by induction that $U(\mathbb{Z}_{2^r})$ is not cyclic for $r \geq 3$.

10. (a) Use Proposition 26.7 to find the remainder when 37^{98} is divided by 24.

(b) True or false: if a, n are coprime positive integers and $a^2 \equiv 1 \bmod n$, then $a \equiv \pm 1 \bmod n$.

(c) True or false: if a, m, n are positive integers such that $\mathrm{hcf}(a, n) = \mathrm{hcf}(m, \phi(n)) = 1$, then

$$a^m \equiv 1 \bmod n \Rightarrow a \equiv 1 \bmod n.$$

11. Find the smallest possible size of a group that contains elements of each of the orders $1, 2, 3, \ldots, 10$, and give an example of such a group.

12. (a) Prove that if $2^p - 1$ is prime, then $2^{p-1}(2^p - 1)$ is a perfect number.

(b) Show that $2^{13} - 1$ and $2^{17} - 1$ are both prime.

(c) Write down the first six even perfect numbers.

13. Find all prime factors less than 100 of the numbers $2^{23} - 1$, $3^{13} - 1$ and $79^{11} - 1$.

14. Let n be a positive integer that is coprime to 10. Prove that the period of the decimal expansion of $\frac{1}{n}$ divides $\phi(n)$.

15. Can you find a non-prime value of n for which the period of $\frac{1}{n}$ is equal to $\phi(n)$?

16. Let G be a group and let H be a subgroup of G.

(a) Prove that the relation $\sim$ on G defined by

$$x \sim y \Leftrightarrow xy^{-1} \in H \ (x, y \in G)$$

is an equivalence relation.

(b) Show that the equivalence class of $\sim$ containing the element x is equal to the right coset Hx.

(c) Deduce that for $x, y \in G$,

$$Hx = Hy \Leftrightarrow xy^{-1} \in H.$$

(d) Now let $G = S_4$ and let H be the cyclic subgroup $\langle (1234) \rangle$ of G. Write down all the distinct right cosets of H in G.

17. Let G be a finite group with a subgroup H, and let $r = |G|/|H|$. Prove that for any $x \in G$ there is an integer $k \in \{1, \ldots, r\}$ such that $x^r \in H$.

18. Critic Ivor Smallbrain, along with great film directors Michael Loser, Ally Wooden and several others, has been invited to be on the judging panel at an all-night viewing of the films which are candidates for the fabulous award known in the profession as the OSCAG (the Order of the Symmetric, Cyclic and Alternating Groupies). There is not enough time for the whole panel to see all the films, so to make things fair they agree to operate according to the following strict rules:

   (1) each film is watched by exactly three panel members;

   (2) for any two members of the panel, there is exactly one film that they both watch;

   (3) for any two films, there is exactly one panel member who watches both of them.

   How many panel members are there? What is the number of candidate films? And how many of them does Ivor get to watch?

Solutions to Odd-Numbered Exercises

Chapter 1

1. TTFTFFTFT

3. (a) Not valid.

(b) Valid. Let C be the statement "I eat chocolate" and D the statement "I am depressed." We are given $C \Rightarrow D$. Therefore $\bar{D} \Rightarrow \bar{C}$. We are told that $\bar{D}$ is true, so I am not eating chocolate. (Actually I am; it's delicious.)

(c) Valid. Let E be the statement "movie was made in England," W be "movie is worth seeing," I be "movie reviewed by Ivor." Then we are given $\bar{W} \Rightarrow \bar{E}$ and $W \Rightarrow I$. Hence $\bar{I} \Rightarrow \bar{W} \Rightarrow \bar{E}$. Therefore, the movie was not made in England.

5. (a) True

(b) False

(c) False

(d) False: for example, if $a = b = 2$ then ab is a square but a, b are not.

(e) True: if a, b are squares then $a = m^2, b = n^2$ with m, n integers, so $ab = m^2 n^2 = (mn)^2$, a square.

7. We produce counterexamples to each of these statements:

(a) $2^4 - 2 = 14$ is not divisible by 4. Hence the statement is false when $n = 2, k = 4$.

(b) 7 is not the sum of three squares.

9. (a) Negation: $\exists n \in \mathbb{Z}$ such that n is a prime number and n is even. This negation is true, as $n = 2$ is an even prime.

(b) Negation: $\exists n \in \mathbb{Z}$ such that $\forall a,b,c,d,e,f,g,h, \in \mathbb{Z}$,

$$n \neq a^3 + b^3 + c^3 + d^3 + e^3 + f^3 + g^3 + h^3.$$

The original statement is true — i.e., every integer is the sum of 8 cubes. This is tricky to prove. Here is a proof.

Let n be any integer. By Q6(c), $n^3 - n$ is a multiple of 6. Write $n^3 - n = 6x$ with $x \in \mathbb{Z}$. Observe that $(x+1)^3 + (x-1)^3 = 2x^3 + 6x$, and hence

$$n = n^3 - 6x = n^3 - (x+1)^3 - (x-1)^3 + 2x^3.$$

Thus n is in fact the sum of 5 cubes.

(Notice that it is crucial in this proof to allow negative cubes. If we insist on all of the cubes $a^3, \ldots, h^3$ being *non-negative* (i.e., $a \geq 0, \ldots, h \geq 0$), then the negation is true — for example, 23 is not the sum of 8 non-negative cubes.)

(c) Negation: $\forall x \in \mathbb{Z}$, $\exists n \in \mathbb{Z}$ such that $x = n^2 + 2$. This negation is false; for example, $x = 4$ is not of the form $n^2 + 2$. So the original statement is true.

(d) Negation: $\forall x \in \mathbb{Z}$, $\exists n \in \mathbb{Z}$ such that $x = n + 2$. This negation is true: take $n = x - 2$.

(e) Negation: $\exists y \in \{x \mid x \in \mathbb{Z}, x \geq 1\}$ such that $5y^2 + 5y + 1$ is not prime. This negation is true: e.g., if $y = 12$ then $5y^2 + 5y + 1 = 781$, which is not prime as it is a multiple of 11.

(f) Negation: $\exists y \in \{x \mid x \in \mathbb{Z}, x^2 < 0\}$ such that $5y^2 + 5y + 1$ is not prime. This negation is false, as the set in question is the empty set. So the original statement is true.

11. (a) Let $x = a_1$. Then $a_x = 3$. Since $a_x > a_{x-1} > \ldots > a_1 \geq 1$, it follows that $x \leq 3$. Easily see that only $x = 2$ is possible. Thus $a_1 = 2$.

(b) Since $a_1 = 2$, have $a_2 = a_{a_1} = 3$, and then $a_3 = a_{a_2} = 6$. Therefore, $a_6 = a_{a_3} = 9$. As $a_3 < a_4 < a_5 < a_6$, it follows that $a_4 = 7, a_5 = 8$. Then $a_7 = a_{a_4} = 12$, $a_8 = 15$, $a_9 = 18$.

(c) Work as in (b) to see that $a_{10}, a_{11}, \ldots$ is

$$19, 20, 21, 22, 23, 24, 25, 26, 27,$$
$$30, 33, 36, 39, 42, 45, 48, 51, 54,$$
$$55, 56, \ldots\ldots, 81,$$
$$84, 87, \ldots, 162,$$
$$163, 164, \ldots, 243,$$
$$246, 249, \ldots\ldots$$

and so on. In particular, $a_{100} = 181$.

Chapter 2

1. (a) By contradiction. Suppose $\sqrt{3}$ is rational, so $\sqrt{3} = \frac{m}{n}$ where m, n are integers and $\frac{m}{n}$ is in lowest terms. Squaring, we get $m^2 = 3n^2$. Thus m^2 is a multiple of 3, and so by Example 1.3, m is a multiple of 3. This means $m = 3k$ for some integer k. Then $3n^2 = m^2 = 9k^2$, so $n^2 = 3k^2$. Therefore n^2 is a multiple of 3, hence so is n. We have now shown that both m and n are multiples of 3. But $\frac{m}{n}$ is in lowest terms, so this is a contradiction. Therefore $\sqrt{3}$ is irrational.

(b) By contradiction again. Suppose $\sqrt{3} = r + s\sqrt{2}$ with r, s rational. Squaring, we get $3 = r^2 + 2s^2 + 2rs\sqrt{2}$. If $rs \neq 0$ this gives $\sqrt{2} = \frac{3 - r^2 - 2s^2}{2rs}$. Since r, s are rational, this implies that $\sqrt{2}$ is rational, which is a contradiction. Hence $rs = 0$. If $s = 0$ then $r^2 = 3$, so $r = \sqrt{3}$, contradicting the fact that $\sqrt{3}$ is irrational by (a). Therefore, $r = 0$ and $3 = 2s^2$. Writing $s = \frac{m}{n}$ in lowest terms, we have $3n^2 = 2m^2$. Now the proof of Proposition 2.3 shows that m and n must both be even, which is a contradiction.

3. (a) True: if $x = m/n$ and $y = p/q$ are rational, so is $xy = mp/nq$.

(b) False: for a counterexample take the irrationals $\sqrt{2}$ and $-\sqrt{2}$. Their product is -2, which is rational.

(c) False: the product of the two irrationals $\sqrt{2}$ and $1 + \sqrt{2}$ is $\sqrt{2} + 2$, which is irrational.

(d) True: we prove it by contradiction. Suppose there is a rational $a \neq 0$ and an irrational b such that $c = ab$ is rational. Then $b = \frac{c}{a}$, and since a and c are rational, this implies that b is rational, a contradiction.

5. By contradiction. Let $\alpha = \sqrt{2} + \sqrt{n}$, and suppose α is rational. Then $\alpha - \sqrt{2} = \sqrt{n}$. Squaring both sides, $\alpha^2 + 2 - 2\alpha\sqrt{2} = n$, so $2\alpha\sqrt{2} = \alpha^2 + 2 - n$. Since clearly $\alpha \neq 0$, we can divide through by 2α to get $\sqrt{2} = (\alpha^2 + 2 - n)/2\alpha$. As α is rational, this implies that $\sqrt{2}$ is rational, which is a contradiction. Hence α is irrational.

7. Let $a_n = \sqrt{n-2} + \sqrt{n+2}$. Then $a_n^2 = (n-2) + (n+2) + 2\sqrt{(n-2)(n+2)} = 2n + 2\sqrt{n^2 - 4}$. We are given that a_n is an integer. This implies $\sqrt{n^2 - 4}$ is rational. By the hint given, this means that $n^2 - 4$ must be a perfect square; i.e., $n^2 - 4 = m^2$ for some integer m. Then $n^2 - m^2 = 4$. Staring at the list of squares $0, 1, 4, 9, 16, \ldots$, we see that the only way the two squares n^2, m^2 can differ by 4 is to have $n^2 = 4, m^2 = 0$. Hence $n = 2$, so $a_n = 2$.

Chapter 3

1. $\frac{1812}{999}$

3. (a) This number is $0.12340123401234....$ which is periodic, hence rational.

(b) Suppose the given number is rational. Then the decimal must be periodic, so for some n there is a sequence $a_1a_2...a_n$ of n digits which repeats itself. But if we go far enough along the decimal, there will be a sequence of at least $2n$ consecutive zeroes. The sequence $a_1a_2...a_n$ must appear within this sequence of zeroes, hence must be just $000...0$. This means the decimal ends $0000....$ (going on forever), which is plainly false. Hence (by contradiction), the number is irrational.

(c) This is $1.1001000010000001.....$ There are ever-increasing sequences of zeroes, so the argument for (b) shows that this number is irrational.

5. Let $x = \frac{100}{9899}$. It's easy to check that $100 + x + 100x = 10000x$. Writing this equation out in full, using the fact that $x = 0.010102...$, and spacing it out a bit to make the point, we get

$$
\begin{array}{l}
\quad\quad 100 \\
+ \quad\ 0\,.\,01\ 01\ 02\ 03\ 05\ ... \\
+ \quad\ 1\,.\,01\ 02\ 03\ 05\ 08\ ... \\
= \\
\quad\quad 101\,.\,02\ 03\ 05\ 08\ 13\ ...
\end{array}
$$

You can see that, at least near the beginning of the decimal expansion of x, the numbers formed by looking at successive pairs of digits are being forced to obey the rule defining the Fibonacci sequence. Eventually one has to start carrying digits and the whole thing goes wrong, but it does work for a while (in fact up to the term 55 in the Fibonacci sequence).

If you try, for example, $x = 1000/998999$, then it works for even longer.

7. Suppose $x = \frac{m}{n}$ is a rational which has decimal expression ending in repeating zeroes. Say $x = a_0.a_1a_2...a_k000....$ Then $x = \frac{A}{10^k}$, where A is an integer (in fact $A = a_k + 10a_{k-1} + ... + 10^k a_0$). So $x = \frac{A}{2^k 5^k}$. Cancelling out common factors of A and $2^k 5^k$, we see that the denominator n is of the form $2^a 5^b$.

Conversely, if $n = 2^a 5^b$, then $\frac{m}{n}$ is equal to $\frac{p}{10^k}$ for some integers p, k, so the decimal expression for $\frac{m}{n}$ has repeating zeroes from the $k + 1^{th}$ position at least.

Chapter 4

1. $(50)^{3/4}(\frac{5}{\sqrt{2}})^{-1/2} = (2.5^2)^{3/4}(5.2^{-1/2})^{-1/2} = 2^{3/4}5^{3/2}5^{-1/2}2^{1/4} = 2^1 5^1 = 10$

3. 2^{617}; $3^{(3^{332})}$

5. Note that $10000^{100} = (100^2)^{100} = 100^{200}$. So 100^{10000} is bigger.

Suppose $2^{1/2} \geq 3^{1/3}$. Taking sixth powers, this implies $2^3 \geq 3^2$, which is false. Hence $2^{1/2} < 3^{1/3}$.

7. $x + y \geq 2\sqrt{xy} \Leftrightarrow (x+y)^2 \geq 4xy \Leftrightarrow x^2 + y^2 + 2xy \geq 4xy \Leftrightarrow (x-y)^2 \geq 0$, which is true. Equality holds when $x = y$.

9. Taking sixth powers of both sides of the equation $y^{4/3} = x^{5/6}$ gives $y^8 = x^5$. Hence $y = 2^5, x = 2^8$ is a solution. It is the smallest solution, since if $1 < y < 2^5$ and y is an integer, then $x = y^{8/5}$ is not an integer.

So the bill was £512.32. Pretty expensive meal.

Chapter 5

1. If $y < 0$ then $-y > 0$, so $x.(-y) > 0$, hence $-xy > 0$, hence $xy < 0$. If $a > b > 0$ then $a - b > 0$, $\frac{1}{a} > 0$ and $\frac{1}{b} > 0$ (using Example 4.4). Hence the product $(a - b).\frac{1}{a}.\frac{1}{b} > 0$. This means $\frac{a-b}{ab} > 0$, so $\frac{1}{b} - \frac{1}{a} > 0$, so $\frac{1}{b} > \frac{1}{a}$.

3. This translates to $3x^2 - 4x + 1 < 0$, which is $(3x - 1)(x - 1) < 0$, which is true provided $\frac{1}{3} < x < 1$.

5. Inequality translates to $\frac{(u-1)(v-1)}{1+uv} > 0$, which is true if $0 < u, v < 1$. The other ranges for which it is true are: $u, v > 1$; or $u, v < 1$ and $uv > -1$; or $u < 1, v > 1$ and $uv < -1$; or $u > 1, v < 1$ and $uv < -1$.

7. (i) The inequality says either $x + 5 \geq 1$ or $x + 5 \leq -1$; i.e., either $x \geq -4$ or $x \leq -6$.

(ii) The inequality is true if and only if $(x + 5)^2 > (x - 2)^2$, i.e. $x^2 + 10x + 25 > x^2 - 4x + 4$, i.e. $x > -\frac{3}{2}$.

(iii) Note that $x^2 + 2x + 3 = (x + 1)^2 + 2$, which is always positive, so the right-hand side of the inequality is always $x^2 + 2x + 3$. When $x \geq -5$, the inequality is $x^2 + 2x + 3 > x + 5$, i.e. $(x+2)(x-1) > 0$, which is true provided $x > 1$ or $x < -2$. And when $x < -5$ the inequality is $x^2 + 2x + 3 > -x - 5$, i.e., $x^2 + 3x + 8 > 0$, which is always true. Hence the range for which the inequality holds is $x > 1$ or $x < -2$.

9. First apply Example 5.14 taking $x = (a_1 a_2)^{1/2}$, $y = (a_3 a_4)^{1/2}$, to get

$$(a_1 a_2 a_3 a_4)^{1/4} \leq \frac{1}{2}((a_1 a_2)^{1/2} + (a_3 a_4)^{1/2}).$$

By Example 5.14, $(a_1 a_2)^{1/2} \leq \frac{1}{2}(a_1 + a_2)$ and $(a_3 a_4)^{1/2} \leq \frac{1}{2}(a_3 + a_4)$. Applying these to the above inequality gives the result.

To prove the Arithmetic-Geometric Mean Inequality for $n = 8$, just repeat the above argument:

$$
\begin{aligned}
(a_1 a_2 \ldots a_8)^{1/8} &\leq \tfrac{1}{2}((a_1 \ldots a_4)^{1/4} + (a_5 \ldots a_8)^{1/4}) \quad \text{by Example 5.14} \\
&\leq \tfrac{1}{2}(\tfrac{1}{4}(a_1 + \cdots + a_4) + \tfrac{1}{4}(a_5 + \cdots + a_8)) \\
&= \tfrac{1}{8}(a_1 + \cdots + a_8).
\end{aligned}
$$

Repeating this argument we can prove the Arithmetic-Geometric Mean Inequality for any case where n is a power of 2.

11. Squaring both sides of $x + y + z = 0$, we get $x^2 + y^2 + z^2 + 2(xy + yz + xz) = 0$. Hence $xy + yz + xz = -\frac{1}{2}(x^2 + y^2 + z^2) \leq 0$.

Chapter 6

1. (a) Let $u = x + iy$, $v = p + iq$. Then $u + v = x + p + i(y+q) = p + x + i(q+y) = v + u$.

(b) With u, v as in (a), $uv = xp - yq + i(xq + yp) = vu$.

(c,d) Routine, I won't bore you with any more details.

(e) This can be checked directly from the definition of complex multiplication, but here's a more subtle argument which shows "why" it's true. Write $u = re^{i\alpha}$, $v = se^{i\beta}$, $w = te^{i\gamma}$. Then

$$u(vw) = re^{i\alpha}(se^{i\beta}.te^{i\gamma}) = re^{i\alpha}(ste^{i(\beta+\gamma)}) = r(st)e^{i(\alpha+(\beta+\gamma))}.$$

Similarly, we see that $(uv)w = (rs)te^{i((\alpha+\beta)+\gamma)}$. Since $r, s, t, \alpha, \beta, \gamma$ are all real, $r(st) = (rs)t$ and $\alpha + (\beta+\gamma) = (\alpha+\beta)+\gamma$, so the expressions for $u(vw)$ and $(uv)w$ are equal.

3. (a) In polar form, $\sqrt{3} - i = 2e^{-i\pi/6}$. Hence $(\sqrt{3} - i)^{10} = 2^{10}e^{-10i\pi/6} = 2^{10}e^{i\pi/3} = 2^{10}(\cos \pi/3 + i\sin \pi/3) = 2^9(1 + i\sqrt{3})$. Similarly, $(\sqrt{3} - i)^{-7} =$

$2^{-7}e^{7i\pi/6} = 2^{-8}(-\sqrt{3}-i)$. Finally, $(\sqrt{3}-i)^n = 2^n e^{-in\pi/6}$, which is real if and only if n is a multiple of 6.

(b) Since $i = e^{i\pi/2}$, the square roots of i are $e^{i\pi/4}$ and $e^{5i\pi/4}$; these are equal to $\pm\frac{1}{\sqrt{2}}(1+i)$.

(c) The 10th roots of i are $e^{i(\pi/20+2k\pi/10)}$ for $0 \le k < 10$. The one nearest to i is the one with argument closest to $\pi/2$, which corresponds to $k = 2$.

(d) The equation is $z^7 = \sqrt{3} - i = 2e^{-i\pi/6}$. If we write $\alpha = 2^{1/7}e^{-i\pi/42}$ and $\omega = e^{2i\pi/7}$, then the seven roots are $\alpha, \alpha\omega, \alpha\omega^2, \alpha\omega^3, \alpha\omega^4, \alpha\omega^5, \alpha\omega^6$. These are

$$2^{1/7}e^{-i\pi/42},\ 2^{1/7}e^{11i\pi/42},\ 2^{1/7}e^{23i\pi/42},\ 2^{1/7}e^{35i\pi/42},$$

$$2^{1/7}e^{47i\pi/42},\ 2^{1/7}e^{59i\pi/42},\ 2^{1/7}e^{71i\pi/42}.$$

The closest to the imaginary axis is $2^{1/7}e^{23i\pi/42}$.

5. The three cube roots of unity $1, \omega$ and ω^2 are the corners of an equilateral triangle (since you can easily check that the distance between any two of these is $\sqrt{3}$). For a general complex number $z \ne 0$, if α is one cube root of z, then the other two are $\alpha\omega$ and $\alpha\omega^2$; hence the sides of the triangle they form all have length $\sqrt{3}|\alpha|$, and so the triangle is again equilateral.

7. (a) If $x = 1$ the equation is true (both sides are 0); and if $x \ne 1$ then $1 + x + x^2 + x^3 + x^4$ is a geometric series with sum $\frac{x^5-1}{x-1}$, from which the equation follows.

Since $\omega^5 = 1$, we have $0 = \omega^5 - 1 = (\omega-1)(\omega^4 + \omega^3 + \omega^2 + \omega + 1)$; since $\omega - 1 \ne 0$, the second factor must be 0.

(b) Notice that $\omega^4 = e^{8\pi i/5} = e^{-2\pi i/5} = \bar\omega$. Hence $\omega + \omega^4 = \omega + \bar\omega = 2\cos\frac{2\pi}{5}$. Similarly $\omega^2 + \omega^3 = 2\cos\frac{4\pi}{5}$.

Observe that $\alpha + \beta = \omega + \omega^2 + \omega^3 + \omega^4 = -1$ and check that $\alpha\beta = -1$ also. Therefore α, β are the roots of the quadratic $(x - \alpha)(x - \beta) = x^2 - (\alpha + \beta)x + \alpha\beta = x^2 + x - 1 = 0$. The roots of this are $\frac{1}{2}(-1\pm\sqrt{5})$. As α is positive, we deduce $\alpha = \frac{1}{2}(-1+\sqrt{5})$, giving the answer.

9. The fact that $|z| = 1$ implies that $z = \cos\theta + i\sin\theta$ for some θ. As $|\sqrt{2} + z| = 1$, this gives $(\cos\theta + \sqrt{2})^2 + \sin^2\theta = 1$, hence $2\sqrt{2}\cos\theta + 2 = 0$, hence $\cos\theta = -\frac{1}{\sqrt{2}}$. Therefore, $\theta = 3\pi/4$ or $5\pi/4$ and the solutions are $z = e^{3i\pi/4}, e^{5i\pi/4}$. For both of these we have $z^8 = 1$.

11. As w is a root of unity, it has modulus 1; hence $w\bar w = |w|^2 = 1$, and so $\bar w = \frac{1}{w}$. It follows that

$$\overline{(1-\omega)}^n = (1-\bar\omega)^n = (1 - \frac{1}{\omega})^n = \frac{1}{\omega^n}(\omega - 1)^n = (\omega - 1)^n.$$

Finally, $\overline{(1-\omega)}^{2n} = ((\overline{1-\omega})^n)^2 = ((\omega-1)^n)^2 = (\omega-1)^{2n} = (1-\omega)^{2n}$, and hence $(1-\omega)^{2n}$ is real.

13. Observe that $|z| = |z+1| = 1$. As $|z| = 1$, we know that $z = \cos\theta + i\sin\theta$ for some θ. Since $|z+1| = 1$, it follows that $(1+\cos\theta)^2 + \sin^2\theta = 1$. This gives $1 = 2 + 2\cos\theta$, so $\cos\theta = -1/2$. Hence $\theta = 2\pi/3$ or $4\pi/3$. So Tantrum is $e^{2i\pi/3}$ or $e^{4i\pi/3}$ and Overthetop is correspondingly $1 + e^{2i\pi/3} = e^{i\pi/3}$ or $1 + e^{4i\pi/3} = e^{-i\pi/3}$. All of these are sixth roots of 1.

Chapter 7

1. (a) Roots are $3+i, 2-i$.

(b) Since $x^4 + x^2 + 1 = (x^6 - 1)/(x^2 - 1)$, the roots of $x^4 + x^2 + 1 = 0$ are the sixth roots of unity, excluding ± 1. These are $e^{\pm i\pi/3}, e^{\pm 2i\pi/3}$.

(c) Since $1+i$ is a root, and the given quartic equation has real coefficients, the conjugate $1-i$ must also be a root. So $(x-(1+i))(x-(1-i)) = x^2 - 2x + 2$ is a factor of the quartic; indeed the equation is $(x^2 - 2x + 2)(2x^2 - 1) = 0$. So the other roots are $\pm 1/\sqrt{2}$.

3. The formula gives roots

$$x = \sqrt[3]{2+11i} + \sqrt[3]{2-11i}.$$

Since a cube root of $2 \pm 11i$ is $2 \pm i$, the roots are

$$(2+i) + (2-i) = 4,$$
$$(2+i)\omega + (2-i)\omega^2 = -2 + \sqrt{3},$$
$$(2+i)\omega^2 + (2-i)\omega = -2 - \sqrt{3}.$$

If we write $2 + 11i = re^{i\theta}$, then $\theta = \tan^{-1}(11/2)$, and the cube root $2 + i = r^{1/3}e^{i\theta/3}$, so $\cos(\theta/3) = 2/\sqrt{5}$.

5. By Example 6.4 we have $\cos 3\theta = 4\cos^3\theta - 3\cos\theta$. Putting $\theta = 2\pi/9$, $c = \cos\theta$, we get $-\frac{1}{2} = 4c^3 - 3c$, hence c is a root of the cubic equation $8x^3 - 6x + 1 = 0$. The other roots are $\cos 4\pi/9$ and $\cos 8\pi/9$. By Proposition 7.1, the sum of these three roots is $-1/8$ times the coefficient of x^2, which is 0; and the product of the roots is $-1/8$ times the constant term, which is $-1/8$.

7. (a) Let the roots be $\alpha, \alpha + d, \alpha + 2d$. Then

$$3\alpha + 3d = -6,$$
$$\alpha(\alpha+d) + \alpha(\alpha+2d) + (\alpha+d)(\alpha+2d) = k,$$
$$\alpha(\alpha+d)(\alpha+2d) = 10.$$

The first equation gives $\alpha + d = -2$, whence the last gives $(-2-d)(-2)(-2+d) = 10$. Hence $d^2 = 9$, so either $d = 3, \alpha = -5$ or $d = -3, \alpha = 1$. Both possibilities give roots $1, -2, -5$ and $k = 3$.

(b) We have $\alpha + \beta + \gamma = 0$, $\alpha\beta + \alpha\gamma + \beta\gamma = -1$, $\alpha\beta\gamma = -1$. Hence

$$\alpha^2 + \beta^2 + \gamma^2 = (\alpha + \beta + \gamma)^2 - 2(\alpha\beta + \alpha\gamma + \beta\gamma) = 0 - (-2) = 2,$$
$$\alpha^2\beta^2 + \alpha^2\gamma^2 + \beta^2\gamma^2 = (\alpha\beta + \alpha\gamma + \beta\gamma)^2 - 2(\alpha^2\beta\gamma + \alpha\beta^2\gamma + \alpha\beta\gamma^2)$$
$$= 1 - 2\alpha\beta\gamma(\alpha + \beta + \gamma) = 1,$$
$$\alpha^2\beta^2\gamma^2 = 1.$$

Hence the cubic with roots $\alpha^2, \beta^2, \gamma^2$ is $x^3 - 2x^2 + x - 1 = 0$. Similarly,

$$\frac{1}{\alpha} + \frac{1}{\beta} + \frac{1}{\gamma} = \frac{\alpha\beta + \alpha\gamma + \beta\gamma}{\alpha\beta\gamma} = 1,$$
$$\frac{1}{\alpha\beta} + \frac{1}{\alpha\gamma} + \frac{1}{\beta\gamma} = \frac{\alpha + \beta + \gamma}{\alpha\beta\gamma} = 0,$$
$$\frac{1}{\alpha\beta\gamma} = -1,$$

so cubic with roots $\frac{1}{\alpha}, \frac{1}{\beta}, \frac{1}{\gamma}$ is $x^3 - x^2 + 1 = 0$.

(c) Let the roots be α, β, γ, where $\alpha + \beta = 1$. We have $\alpha + \beta + \gamma = -p$, whence $\gamma = -p - 1$. Next, $p^2 = \alpha\beta + \alpha\gamma + \beta\gamma = \gamma(\alpha + \beta) + \alpha\beta$, whence $\alpha\beta = p^2 - \gamma(\alpha + \beta) = p^2 + p + 1$. Finally, $\alpha\beta\gamma = -r$ gives $(p^2 + p + 1)(-p - 1) = -r$.

(d) Let the roots be $\alpha, \beta, \gamma, \delta$, where $\alpha + \beta = 5$. We have $\alpha + \beta + \gamma + \delta = 3$, so $\gamma + \delta = -2$. Next, the sum of products of pairs of roots is -5, so $(\alpha + \beta)(\gamma + \delta) + \alpha\beta + \gamma\delta = -5$, whence $\alpha\beta + \gamma\delta = 5$. Also, $\alpha\beta\gamma\delta = -6$. Thus $\alpha\beta, \gamma\delta$ are the roots of the quadratic $x^2 - 5x - 6 = 0$, which are $6, -1$. If $\alpha\beta = -1, \gamma\delta = 6$, then α, β are the roots of $x^2 - 5x - 1 = 0$ and γ, δ are the roots of $x^2 + 2x + 6 = 0$, so the original quartic must factorise as $(x^2 - 5x - 1)(x^2 + 2x + 6)$; but the x coefficient of this product is not 17, so this is not the case. Hence $\alpha\beta = 6, \gamma\delta = -1$, and α, β are the roots of $x^2 - 5x + 6 = 0$ and γ, δ are the roots of $x^2 + 2x - 1 = 0$. Thus the roots of the quartic are $3, 2, -1 + \sqrt{2}$ and $-1 - \sqrt{2}$.

Chapter 8

1. For $n \geq 8$ let $P(n)$ be the statement that it is possible to pay n roubles using only 3 and 5 rouble notes. Then $P(8)$ is true. Assume $P(n)$ true. Then $n = 3a + 5b$, where a is the number of 3 rouble notes and b the number of 5 rouble notes. Since $n \geq 8$, either $b \geq 1$ or $a \geq 3$ (or both). If $b \geq 1$, to pay $n + 1$ roubles, use $a + 2$ 3's and $b - 1$ 5's; and if $a \geq 3$ use $a - 3$ 3's and $b + 2$ 5's. Hence $P(n) \Rightarrow P(n+1)$.

3. (a) The formula is $\sum_{r=1}^{n} r^3 = (\frac{1}{2}n(n+1))^2$. The induction proof is the usual kind of thing.

(b) General formula is $\sum_{r=n^2+1}^{(n+1)^2} r = n^3 + (n+1)^3$. To prove it, observe that the left-hand side is $\sum_{r=1}^{(n+1)^2} r - \sum_{r=1}^{n^2} r = \frac{1}{2}(n+1)^2((n+1)^2+1) - \frac{1}{2}n^2(n^2+1)$. This works out to be $n^3 + (n+1)^3$.

5. (a) Let $P(n)$ be the statement that $5^{2n} - 3^n$ is divisible by 11. Then $P(0)$ is true. Suppose $P(n)$ is true, so $5^{2n} - 3^n = 11k$ for some integer k. Now

$$5^{2(n+1)} - 3^{n+1} = 25 \cdot 5^{2n} - 3 \cdot 3^n = 22 \cdot 5^{2n} + 3(5^{2n} - 3^n) = 22 \cdot 5^{2n} + 33k,$$

which is divisible by 11. Hence $P(n) \Rightarrow P(n+1)$. Therefore $P(n)$ is true for all $n \geq 0$ by induction.

(b) Let $P(n)$ be the statement that 2^{4n-1} ends with an 8; i.e., $2^{4n-1} = 10k + 8$ for some integer k. Clearly $P(1)$ is true, as $2^{4-1} = 8$. Assume $P(n)$ is true, say $2^{4n-1} = 10k + 8$. Then $2^{4(n+1)-1} = 2^4 \cdot 2^{4n-1} = 16(10k+8) = 160k + 128$, which ends with an 8. Hence $P(n) \Rightarrow P(n+1)$. Therefore $P(n)$ is true for all $n \geq 1$ by induction.

(c) Let $P(n)$ be the statement that $n^3 + (n+1)^3 + (n+2)^3$ is divisible by 9. Then $P(1)$ is true as $1^3 + 2^3 + 3^3 = 36$. Assume $P(n)$ is true, so $n^3 + (n+1)^3 + (n+2)^3 = 9k$ for some integer k. Now $(n+1)^3 + (n+2)^3 + (n+3)^3 = (n^3 + (n+1)^3 + (n+2)^3) + ((n+3)^3 - n^3) = 9k + (9n^2 + 27n + 27)$, which is clearly divisible by 9. Hence $P(n) \Rightarrow P(n+1)$.

(d) Let $x \geq 2$. Let $P(n)$ be the statement that $x^n \geq nx$. Then $P(1)$ says that $x \geq x$, which is true. Assume $P(n)$. Multiplying both sides by x (as we may do, since x is positive), we have $x^{n+1} \geq nx^2$. Observe that $nx^2 \geq (n+1)x$ since this is equivalent to $x \geq \frac{n+1}{n}$, which is true as $x \geq 2$. Hence $x^{n+1} \geq nx^2 \geq (n+1)x$, and so we have shown that $P(n) \Rightarrow P(n+1)$.

(e) Let $P(n)$ be the statement $5^n > 4^n + 3^n + 2^n$. Check that $P(3)$ is true. Assume $P(n)$ and multiply both sides by 5. This gives

$$5^{n+1} > 5(4^n + 3^n + 2^n) = 5 \cdot 4^n + 5 \cdot 3^n + 5 \cdot 2^n$$
$$> 4 \cdot 4^n + 3 \cdot 3^n + 2 \cdot 2^n = 4^{n+1} + 3^{n+1} + 2^{n+1}.$$

Hence $P(n) \Rightarrow P(n+1)$.

7. Let $P(n)$ be the statement that $f_n = \frac{1}{\sqrt{5}}(\alpha^n - \beta^n)$. Since $\frac{1}{\sqrt{5}}(\alpha - \beta) = 1 = f_1$, $P(1)$ is true. Now assume $P(1), \ldots, P(n)$ are all true. Then

$$f_{n+1} = f_n + f_{n-1} = \frac{1}{\sqrt{5}}(\alpha^n - \beta^n) + \frac{1}{\sqrt{5}}(\alpha^{n-1} - \beta^{n-1})$$
$$= \frac{1}{\sqrt{5}}(\alpha^{n-1}(\alpha+1) - \beta^{n-1}(\beta+1)).$$

Now $\alpha + \beta = 1$ and $\alpha\beta = -1$, so α, β are the roots of the quadratic $x^2 - x - 1 = 0$, and so $\alpha^2 = \alpha + 1, \beta^2 = \beta + 1$. It follows that $f_{n+1} = \frac{1}{\sqrt{5}}(\alpha^{n+1} - \beta^{n+1})$, which is $P(n+1)$. Hence the result by strong induction.

9. Let $P(n)$ be the statement $(1+q)^n \leq 1+2^n q$. Clearly $P(1)$ is true. Assume $P(n)$ true and multiply both sides by the positive number $1+q$. Then

$$
\begin{aligned}
(1+q)^{n+1} &\leq (1+2^n q)(1+q) = 1 + 2^n q + q + 2^n q^2 \\
&\leq 1 + 2^n q + q + 2^{n-1} q \quad (\text{as } q \leq \tfrac{1}{2}) \\
&= 1 + q(2^n + 1 + 2^{n-1}) \leq 1 + q(2^n + 2^n) = 1 + 2^{n+1} q.
\end{aligned}
$$

Hence $P(n+1)$ is true. Thus $P(n) \Rightarrow P(n+1)$, so $P(n)$ is true for all n by induction.

11. For n a positive integer, let $P(n)$ be the statement that n is a sum of different primes. Certainly $P(1)$ is true (as we are regarding 1 as a prime in this question). Assume $P(1), P(2), \ldots, P(n)$ are all true. By the result stated in the question, there is a prime p such that $\frac{n+1}{2} < p < n+1$. Let $x = n+1-p$. Then $n+1 = p+x$, where $x < p$. By assumption, $P(x)$ is true, so $x = p_1 + \ldots + p_k$, a sum of different primes p_i. Since $x < p$, each p_i is also different from p. Hence $n+1 = p + p_1 + \ldots + p_k$ is a sum of different primes. Thus $P(n+1)$ is true. Hence $P(n)$ is true for all n by strong induction.

13. (a) The answer is $2n$ regions. Let $P(n)$ be the statement that n lines through a single point divide the plane into $2n$ regions. Clearly $P(1)$ is true. Assume $P(n)$ true and consider $n+1$ lines, all through a single point. If we remove 1 of these lines we have n lines; by assumption these divide the plane into $2n$ regions. The extra line goes through two of these regions, dividing each into two; so the $n+1$ lines give 2 more regions than the n lines, hence give $2n+2$ regions. Hence $P(n) \Rightarrow P(n+1)$. So $P(n)$ is true for all n by induction.

(b) There are $2n$ infinite regions and the rest are finite. Briefly: in proving $P(n) \Rightarrow P(n+1)$, note that the $n+1^{th}$ line creates just two new infinite regions. So if there are $2n$ infinite regions with n lines then there are $2n+2$ with $n+1$ lines.

15. (a) Apply the Cauchy inequality, taking $a_1, a_2, a_3 = a, b, c$ and $b_1, b_2, b_3 = b, c, a$.

(b) (i) Example 8.11 shows that $1 = x+y+z \leq \sqrt{3}\sqrt{x^2+y^2+z^2}$, hence $x^2 + y^2 + z^2 \geq \frac{1}{3}$.

(ii) Apply Example 8.12, taking $n = 3$ and $p_1 = p_2 = p_3 = \frac{1}{3}$.

17. (i) By Example 8.11, $\sum_1^n p_i^2 \geq \frac{1}{n}$.

(ii) Applying Cauchy's Inequality to the sequences $\sqrt{p_1}, \ldots, \sqrt{p_n}$ and $\frac{1}{\sqrt{p_1}}, \ldots, \frac{1}{\sqrt{p_n}}$, we get $\sum_1^n \frac{1}{p_i} \geq n^2$.

(iii) Applying Example 8.11 to the sequence $\frac{1}{p_1}, \ldots, \frac{1}{p_n}$ gives

$$
\left(\sum_1^n \frac{1}{p_i} \right)^2 \leq n \sum_1^n \frac{1}{p_i^2},
$$

and so $\sum_1^n \frac{1}{p_i^2} \geq n^3$.

(iv) First expand the sum:

$$\sum_1^n \left(p_i + \frac{1}{p_i} \right)^2 = 2n + \sum_1^n p_i^2 + \sum_1^n \frac{1}{p_i^2}.$$

Hence by the previous parts,

$$\sum_1^n \left(p_i + \frac{1}{p_i} \right)^2 \geq 2n + \frac{1}{n} + n^3 = \frac{1}{n}(n^2 + 1)^2.$$

Chapter 9

1. (a) The proof of (9.2) in the text shows that $3V = 2E$. Obviously $m + n = F$. Now count the pairs e, f, where e is an edge, f is a face and e lies on f. As in the proof of (9.1) in the text, the number of such pairs is $2E$. When f is a square there are 4 possibilities for e, and when f is a pentagon there are 5; hence the total number of pairs is also $4m + 5n$, and so $2E = 4m + 5n$.

(b) Using the first part, Euler's formula $V - E + F = 2$ gives

$$\frac{4m + 5n}{3} - \frac{4m + 5n}{2} + m + n = 2,$$

which works out as $2m + n = 12$.

(c) I can think only of a cube ($m = 6, n = 0$), a dodecahedron ($m = 0, n = 12$) and a 5-sided prism with squares as sides and pentagonal top and bottom ($m = 5, n = 2$).

3. This question is quite tricky. For the proof, we will count the "outside" of the graph as a further face (called the "infinite" face). Counting this, the number of faces is increased by 1, and Theorem 9.2 becomes $v - e + f = 2$. As in the solution to the previous question, counting edge–face pairs gives $2e \geq 3f$. Substituting $f = 2 - v + e$, we get $2e \geq 3(2 - v + e)$, which leads to $e \leq 3v - 6$.

5. As in the hint, suppose every vertex is joined to at least 2 others. This says $v(x) \geq 2$ for all vertices x, where $v(x)$ is the number of vertices joined to x. We'll show there must be at least one face. Pick any vertex x_0 and define a path of edges as follows. Let x_1 be one of the vertices joined to x_0. Since $v(x_1) \geq 2$, there is another vertex x_2 joined to x_1 with $x_2 \neq x_0$. Similarly, since $v(x_2) \geq 2$, there is another vertex x_3 joined to x_2 with $x_3 \neq x_1$ (but x_3 could equal x_0). Carry on like this: we get a path $x_0, x_1, x_2, \ldots$, each vertex joined to the previous one and the next one. As the graph has only a finite number of vertices, eventually one of the x_i must equal a previous x_j. By the way we have chosen the x's, j is not $i-2$. So part of the path is the sequence $x_j, x_{j+1}, \ldots, x_{i-1}, x_j$, which has at least three edges. This is what we call a *circuit* in the graph, and clearly it encloses a face. Hence there is at least one face.

7. It's easy to draw K_4 as a plane graph — just put three vertices at the corners of a triangle and the fourth in the middle of the triangle.

Observe that K_5 has $v = 5$ vertices and $e = 10$ edges. Hence it does not satisfy the inequality $e \leq 3v - 6$, so it cannot be plane by Exercise 3.

9. Let a be the number of pentagons and b the number of hexagons. As in Exercise 1(a), we get

$$3V = 2E, \quad 5a + 6b = 2E, \quad a + b = F.$$

Now consider Euler's formula: $V - E + F = 2$. Multiply through by 6 to get $6V - 6E + 6F = 12$. By the above we know

$$6E = 15a + 18b, \quad 6V = 4E = 10a + 12b, \quad 6F = 6(a+b).$$

Substituting in the equation $6V - 6E + 6F = 12$ gives

$$(10a + 12b) - (15a + 18b) + 6(a+b) = 12.$$

Amazingly, this works out as $a = 12$.

Chapter 10

1. (i) Here $d = \text{hcf}(a,b) = 1$, and $1 = 12a - 7b$.
 (ii) Here $d = 23 = -9a + 7b$.
 (iii) $d = 23 = -6a + 7b$.

3. Observe that $\text{hcf}(7,24) = 1$. Therefore, there are integers s, t such that $1 = 24s + 7t$. Then $7t = 1 - 24s$, so if the train leaves at k o'clock on a particular day, then t trains later the departure time is $k + 1$ o'clock. Hence on some day, the departure time will be 9 o'clock.

Olga sees Ivan every n days, where n is the least positive integer such that $24n$ is a mutiple of 7. Clearly $n = 7$; so she sees him once a week.

If some Vladivostock train leaves at 12 noon then at any *even* hour there is a train some day (as $\text{hcf}(24, 14) = 2$), and once a week the train leaves at 12 noon. Similarly, if some train leaves at 11 a.m. then at any odd hour there is a train some day.

5. (a) Write $a = cm$ with c an integer. Since m, n are coprime, there are integers r, s such that $rm + sn = 1$. Multiplying through by c gives $rmc + snc = c$, hence $ra + snc = c$. Since n divides a, it follows from this equation that n also divides c. Writing $c = en$ we then have $a = emn$. Thus mn divides a.

(b) Suppose $\mathrm{hcf}(m, n) = d > 1$. Write $m = xd, n = yd$ where x, y are integers. Then xyd is divisible by m and n, but not by mn.

(c) Since $\mathrm{hcf}(x, m) = 1$ and $\mathrm{hcf}(y, m) = 1$, there are integers s, t, u, v such that $sx + tm = 1$, $uy + vm = 1$. Then $(sx + tm)(uy + vm) = 1$, so $suxy + (sxv + tuy + tvm)m = 1$. Therefore any common factor of xy and m also divides 1, so $\mathrm{hcf}(xy, m) = 1$.

7. Easy: after n days, where n is the lcm of 3, 4 and 7 — i.e., in 12 weeks.

9. By Proposition 10.3 there are integers x, y such that $1 = xa + yb$. Multiplying through by n, we have $n = nxa + nyb$. So take $s = nx, t = ny$.

Chapter 11

1. $3 \cdot 7 \cdot 11 \cdot 13 \cdot 37$

3. We are given that the prime factorization of n is $n = p_1^{a_1} \cdots p_k^{a_k}$ with all $a_i \geq 2$. We can write each $a_i = 2b_i + 3c_i$ for some positive integers b_i, c_i (for example, if a_i is even take $c_i = 0$, $b_i = a_i/2$ and if a_i is odd take $c_i = 1$, $b_i = \frac{1}{2}(a_i - 3)$). Then $n = x^2 y^3$ where $x = p_1^{b_1} \cdots p_k^{b_k}$ and $y = p_1^{c_1} \cdots p_k^{c_k}$.

5. (a) By contradiction. Suppose $2^{1/3}$ is rational; so $2^{1/3} = \frac{x}{y}$ in lowest terms, where x, y are integers. Cubing gives $x^3 = 2y^3$. Let the prime factorizations of x, y be $x = 2^a p_1^{a_1} \ldots$ and $y = 2^b q_1^{b_1} \ldots$. Then the equation $x^3 = 2y^3$ gives

$$2^{3a} p_1^{3a_1} \ldots = 2^{1+3b} q_1^{3b_1} \ldots$$

By the Fundamental Theorem 11.1, this implies that $3a = 1 + 3b$, which is impossible since a, b are integers. Hence $2^{1/3}$ is irrational.

The same argument shows that $3^{1/3}$ is irrational.

(b) The right-to-left implication is easy: if m is an nth power, then clearly $m^{1/n}$ is rational.

Now for the left-to-right implication. Suppose $m^{1/n}$ is rational; so $m^{1/n} = \frac{x}{y}$, where x, y are integers. Then $x^n = my^n$. Let p be a prime, and let p^a, p^b, p^c

be the largest powers of p which divide x, y, m, respectively. Then the power of p dividing x^n is p^{an}, while the power of p dividing my^n is p^{c+bn}. By the Fundamental Theorem 11.1, we must have $an = c + bn$, and hence $c = n(a - b)$ is divisible by n.

We have shown that the power to which each prime divides m is a multiple of n; in other words, the prime factorization of m is

$$m = p_1^{na_1} \dots p_k^{na_k}$$

for some integers a_i. Hence $m = (p_1^{a_1} \dots p_k^{a_k})^n$, and so m is an nth power, as required.

7. (a) The hcf is $2 \cdot 5^2$ and the lcm is $2^2 \cdot 3 \cdot 5^3$. So the pairs (m, n) are $(2 \cdot 5^2, 2^2 \cdot 3 \cdot 5^3)$, $(2 \cdot 3 \cdot 5^2, 2^2 \cdot 5^3)$, $(2 \cdot 5^3, 2^2 \cdot 3 \cdot 5^2)$, $(2 \cdot 3 \cdot 5^3, 2^2 \cdot 5^2)$.

(b) $hcf(m, n)$ divides m, which divides $lcm(m, n)$; hence $hcf(m, n)$ divides $lcm(m, n)$. They are equal when both equal m, and similarly both equal n, i.e., when $m = n$.

(c) As in Proposition 11.2, let $m = p_1^{r_1} \cdots p_k^{r_k}$, $n = p_1^{s_1} \cdots p_k^{s_k}$. Define x to be the product of all the $p_i^{r_i}$ for which $r_i \geq s_i$, and y to be the product of all the $p_j^{s_j}$ for which $r_j < s_j$.

9. We must show the equation $x^6 - y^5 = 16$ has no solutions $x, y \in \mathbb{Z}$.

Suppose $x, y \in \mathbb{Z}$ are solutions. First suppose x is even. Then y must be even. Hence the LHS of the equation is divisible by 2^5, so it cannot equal 16.

So x must be odd. The equation is $y^5 = x^6 - 16 = (x^3 - 4)(x^3 + 4)$. The hcf of the two factors $x^3 - 4$ and $x^3 + 4$ divides their difference, 8. As both are odd numbers (since x is odd), we deduce that $hcf(x^3 - 4, x^3 + 4) = 1$. So $x^3 - 4, x^3 + 4$ are coprime numbers with product equal to the fifth power y^5. By Proposition 11.4(b), this implies that both $x^3 - 4$ and $x^3 + 4$ are fifth powers. But two fifth powers clearly cannot differ by 8 (the fifth powers are $\dots, -32, -1, 0, 1, 32, \dots$). Hence there are no solutions.

Chapter 12

1. One of the three numbers $p, p + 2, p + 4$ must be divisible by 3. Since they are all supposed to be prime, one of them must therefore be equal to 3, so the only possibility is $p = 3$.

3. For $n = 5, 6, 7, 8, 9, 10$ we have $\phi(n) = 4, 2, 6, 4, 6, 4$, respectively.

If p is prime then all the numbers $1, 2, \dots, p - 1$ are coprime to p, and hence $\phi(p) = p - 1$.

For $r \geq 1$, the numbers between 1 and p^r which are *not* coprime to p^r are those which are divisible by p, namely, the numbers kp with $1 \leq k \leq p^{r-1}$. There are p^{r-1} such numbers, and hence $\phi(p^r) = p^r - p^{r-1}$.

5. $x = 40$ will do nicely.

Chapter 13

1. (a) $7^2 \equiv 5 \bmod 11$, so $7^4 \equiv 5^2 \equiv 3 \bmod 11$ and so $7^5 \equiv 3.7 \equiv -1 \bmod 11$. Therefore $7^{135} \equiv (-1)^{27} \equiv -1 \bmod 11$, so $7^{137} \equiv -7^2 \equiv 6 \bmod 11$. So $r = 6$.

(b) Use the method of successive squares from Example 13.3. Calculate that $2^{16} \equiv 391 \bmod 645$ and $2^{64} \equiv 256 \bmod 645$. Hence $2^{81} = 2^{1+16+64} \equiv 2 \cdot 391 \cdot 256 \equiv 242 \bmod 645$.

(c) We need to consider 3^{124} modulo 100. Observe $3^5 \equiv 43 \bmod 100$, so $3^{10} \equiv 49 \bmod 100$ and then $3^{20} \equiv 1 \bmod 100$. Hence $3^{120} \equiv 1 \bmod 100$, and so $3^{124} \equiv 3^4 \equiv 81 \bmod 100$. Therefore the last two digits of 3^{124} are 81.

(d) The multiple $21n$ will have the last 3 digits 241 if $21n \equiv 241 \bmod 1000$. Since $\mathrm{hcf}(21, 1000) = 1$, such an n exists, by Proposition 13.6.

3. (a) There is a solution by Proposition 13.6, as $\mathrm{hcf}(99, 30) = 3$ divides 18. To find a solution, observe first that $3 = 10 \cdot 30 - 3 \cdot 99$. Multiplying through by 6, we get $18 = 60 \cdot 30 - 18 \cdot 99$, hence $-18 \cdot 99 \equiv 18 \bmod 30$. So $x = -18$ is a solution.

(b) There is no solution by Proposition 13.6, as $\mathrm{hcf}(91, 143) = 13$ does not divide 84.

(c) The squares $0^2, 1^2, 2^2, 3^2, 4^2$ are congruent to $0, 1, 4, 4, 1$ modulo 5, respectively. Since any integer x is congruent to one of $0, 1, 2, 3, 4$ modulo 5, it follows that x^2 is congruent to 0, 1 or 4. Hence the equation $x^2 \equiv 2 \bmod 5$ has no solution.

(d) Putting $x = 0, 1, 2, 3, 4$ gives $x^2 + x + 1$ congruent to $1, 3, 2, 3, 1$ modulo 5, respectively. Hence the the equation $x^2 + x + 1 \equiv 0 \bmod 5$ has no solution.

(e) $x = 2$ is a solution.

5. (a) Since $7 | 1001$, we have $1000 \equiv -1 \bmod 7$, so $1000^2 \equiv 1 \bmod 7$, $1000^3 \equiv -1 \bmod 7$, etc. So the rule is to split the digits of a number n into chunks of size 3 and then alternately add and subtract — then the answer is divisible by 7 if and only if n is. The number 6005004003002001 is congruent modulo 7 to $1 - 2 + 3 - 4 + 5 - 6 = -3$, so the remainder is 4.

(b) Same rule as for 7. The number is again congruent to -3 modulo 13, so the remainder is 10.

(c) Since $1000 \equiv 1 \bmod 37$, $1000^2 \equiv 1 \bmod 37$, etc., the rule is to split the digits of a number n into chunks of size 3 and then add — the answer is divisible by 37 if and only if n is. The given number is congruent modulo 37 to $1+2+3+4+5+6 = 21$.

7. Consider a square n^2. As in Exercise 2(c), $n^2 \equiv 0, 1$ or 4 mod 5. Similarly, we see that $n^2 \equiv 0, 1$ or 4 mod 8.

We first show n is divisible by 5. We know that the squares $2n+1$ and $3n+1$ are congruent to 0,1 or -1 modulo 5. Say $2n+1 \equiv a \bmod 5, 3n+1 \equiv b \bmod 5$, with $a, b \in \{0, 1, -1\}$. If $a \neq b$, then adding gives $5n+2 \equiv 2 \equiv a+b \bmod 5$; but this cannot hold when $a \neq b$ and $a, b \in \{0, 1, -1\}$. So $a = b$; then subtracting gives $n \equiv b - a \bmod 5$; hence as $a = b$, we get $n \equiv 0 \bmod 5$, i.e., n is divisible by 5.

Now we show n is divisible by 8 in exactly the same way. Hence n is divisible by 40.

The first value of n that works is 40, since then $2n+1 = 81$ and $3n+1 = 121$ are squares.

Another value of n that works is 3960, since then $2n+1 = 7921 = 89^2$ and $3n+1 = 11881 = 109^2$.

9. The equation $ax = b$ has a solution for $x \in \mathbb{Z}_p$ if and only if the congruence equation $ax \equiv b \bmod p$ has a solution. Since $a \neq 0$ in $\mathbb{Z}_p$, a and p are coprime, so there is a solution by Proposition 13.6.

11. The number of days in 1000 years is $1000 \times 365 + 250$ (the 250 for the leap years). Since $365 \equiv 1 \bmod 7$, this is congruent to 1250 modulo 7, which is congruent to 4 modulo 7. Hence May 6, 3005 will in fact be a Tuesday.

PS: I have learnt recently that this is wrong, since "century" years (those ending 00) are deemed to be leap years only when divisible by 400. So only the century years 2400 and 2800 are leap years and so the number of leap years in the period is 242 rather than 250. This makes May 6, 3005 a Monday.

Chapter 14

1. (a) By Fermat's Little Theorem, $3^{10} \equiv 1 \bmod 11$, so $3^{301} = 3^{300} \cdot 3 \equiv 3 \bmod 11$. In other words, $3^{301} \pmod{11} = 3$. Likewise, we have $5^{110} \pmod{13} = 12$ and $7^{1388} \pmod{127} = 49$.

(b) By Fermat's Little Theorem, $n^7 \equiv n \bmod 7$. Also $n^3 \equiv n \bmod 3$, and hence $n^7 = n^3 \cdot n^3 \cdot n \equiv n^3 \equiv n \bmod 3$. Clearly also $n^7 \equiv n \bmod 2$. Hence $n^7 - n$ is divisible by 2, 3 and 7, hence by 42, i.e., $n^7 \equiv n \bmod 42$.

3. Let a be coprime to 561. Then by Fermat, $a^{16} \equiv 1 \bmod 17$, $a^{10} \equiv 1 \bmod 11$ and $a^2 \equiv 1 \bmod 3$. So

$$a^{560} \equiv (a^{16})^{35} \equiv 1 \bmod 17,$$
$$a^{560} \equiv (a^{10})^{56} \equiv 1 \bmod 11,$$
$$a^{560} \equiv (a^2)^{280} \equiv 1 \bmod 3,$$

and hence $a^{560} - 1$ is divisible by $3, 11$ and 17, hence by $3 \cdot 11 \cdot 17 = 561$. So $a^{560} \equiv 1 \bmod 561$.

5. By Fermat, $p^{q-1} \equiv 1 \bmod q$. Since $q^{p-1} \equiv 0 \bmod q$, this implies that $p^{q-1} + q^{p-1} \equiv 1 \bmod q$. Similarly, $p^{q-1} + q^{p-1} \equiv 1 \bmod p$. Hence $p^{q-1} + q^{p-1} - 1$ is divisible by both p and q, hence by pq, and so $p^{q-1} + q^{p-1} \equiv 1 \bmod pq$.

7. (a) Use the recipe provided by Proposition 14.2. Since $3 \cdot 19 \equiv 1 \bmod 28$, the solution is $x \equiv 2^{19} \bmod 29$. Using successive squares, this is $x \equiv 26 \bmod 29$.

(b) Notice cleverly that $143 = 11 \cdot 13$, so we use the recipe of Proposition 14.3. Here $(p-1)(q-1) = 120$, and $7 \cdot 103 \equiv 1 \bmod 120$. So the solution is $x \equiv 12^{103} \bmod 143$. Since $12^2 \equiv 1 \bmod 143$, the solution is $x \equiv 12 \bmod 143$.

(c) Again use 14.3. Since $11 \cdot 11 \equiv 1 \bmod 120$, the solution is $2^{11} (\bmod 143)$, which is $46 (\bmod 143)$.

9. Use successive squares to calculate that $2^{1386} \equiv 1 \bmod 1387$, but $2^{693} \equiv 512 \bmod 1387$. So Miller's test shows that 1387 is not prime.

Chapter 15

1. We have $p + q = pq - (p-1)(q-1) + 1 = 18779 - 18480 + 1 = 300$. Hence p, q are the roots of $x^2 - 300x + 18779 = 0$. Using the formula for the roots of a quadratic, these are $\frac{1}{2}(300 \pm \sqrt{300^2 - 4 \cdot 18779})$, i.e., 211 and 89.

3. To crack this code, observe that $1081 = 23 \cdot 47$. Taking $p = 23, q = 47$, we have $(p-1)(q-1) = 1012$. Since $e = 25$ and $25 \cdot 81 \equiv 1 \bmod 1012$, the decoding power $d = 81$. So the decoded message starts with $23^{81} (\bmod 1081) = 161$, then $930^{81} (\bmod 1081) = 925$, then $228^{81} (\bmod 1081) = 30$, and finally $632^{81} (\bmod 1081) = 815$. So the decoded message is 161925030815, which with the usual letter substitutions (A for 01, etc.), is PSYCHO. Good choice, Ivor!

Chapter 16

1. $\binom{8}{3} = \frac{8 \cdot 7 \cdot 6}{3 \cdot 2 \cdot 1} = 56$ and $\binom{15}{5} = 3003$

3. (a) The argument of Example 16.4 shows that the number of solutions to $x+y+z+t = 14$ (x,y,z,t non-negative integers) is equal to the number of choices for the positions of three 0's among 17 symbols, so is equal to $\binom{17}{3}$.

(b) Now let's count the number of solutions with $t \geq 9$. Substituting $u = t-9$, we see that this is the same as the number of solutions to $x+y+z+u=5$ (x,y,z,u non-negative integers), which is $\binom{8}{3}$. Hence the number of solutions to $x+y+z+t = 14$ in non-negative integers with $t \leq 8$ is equal to $\binom{17}{3} - \binom{8}{3} = 624$.

(c) Write $y_i = x - c_i$ for $1 \leq i \leq r$. Then the number of solutions to the given equation is equal to the number of solutions to $y_1 + \cdots + y_r = N - \sum c_i$, where y_i are integers and $y_i \geq 0$ for all i. Each solution can be represented as a string of y_1 1's, then a 0, then y_2 1's, then a 0, and so on; this is a string consisting of $r-1$ 0's and $N - \sum c_i$ 1's. So the number of solutions is equal to the number of choices for the positions of $r-1$ 0's among $N - \sum c_i + r - 1$ symbols, which is

$$\binom{N - \sum c_i + r - 1}{r - 1}.$$

5. (a) The number of words with k letters is 2^k. So the total number with ten or fewer letters is $2 + 2^2 + \cdots + 2^{10} = 2(1 + 2 + \cdots + 2^9) = 2(2^{10} - 1) = 2046$.

(b) There are 3! ways of ordering the letters d, e, f. For a given ordering, there are four spaces before, between and after the letters d, e, f, and a, b, c must be put in different spaces: there are four choices for where to put a, three for b and two for c. So the total number of words is $3! \times 4 \times 3 \times 2 = 144$.

7. (i) Let the wolves be $w_1, \ldots, w_n$. Suppose Liebeck throws the first steak to wolf w_{i_1}, the second to wolf w_{i_2}, and so on. So each possible assignment of steaks corresponds to a sequence $w_{i_1} \ldots w_{i_n}$. The total number of possible sequences is n^n, and the number of sequences for which every wolf gets a steak is the number of arrangements of $w_1, \ldots, w_n$, which is $n!$. Hence the chance that every wolf gets a steak is $\frac{n!}{n^n}$.

(ii) If exactly one wolf does not appear in the sequence $w_{i_1} \ldots w_{i_n}$, then there are two positions having the same wolf. These positions can be chosen in $\binom{n}{2}$ ways, and the repeated wolf in n ways. The remaining $n-2$ positions must be filled by $n-2$ of the remaining $n-1$ wolves, which can be done in $P(n-1, n-2) = (n-1)!$ ways. So the number of sequences in which exactly one wolf is steak-free is equal to $\binom{n}{2}(n-1)!$, and the chance of this event is $\binom{n}{2}(n-1)!/n^n$.

(iii) Presumably Liebeck will be eaten if not all of the wolves get a steak. By (i) the chance of this is $1 - \frac{7!}{7^7}$. This works out as roughly 0.994, or a 99.4% chance of being eaten. Goodbye, Liebeck!

9. Let $P(n)$ be the statement of the Binomial Theorem 16.2. Then $P(1)$ is trivially true as it just says $(a+b)^1 = a+b$. Assume $P(n)$ is true. Multiplying both sides by $a+b$, we get

$$(a+b)^{n+1} = \sum_{r=0}^{n} \binom{n}{r} a^{n-r} b^r (a+b) = \sum_{r=0}^{n} \binom{n}{r} (a^{n-r+1} b^r + a^{n-r} b^{r+1}).$$

For $0 \le s \le n+1$, the coefficient of $a^{n-s} b^s$ in this expression is $\binom{n}{s-1} + \binom{n}{s}$, which by Question 8 is equal to $\binom{n+1}{s}$. Hence

$$(a+b)^{n+1} = \sum_{s=0}^{n+1} \binom{n+1}{s} a^{n-s} b^s,$$

which is $P(n+1)$. This completes the proof by induction.

11. (a) There is an arithmetic progression $a, *, b$ if and only if $b - a$ is even (in which case the triple is $a, \frac{a+b}{2}, b$), i.e., if and only if either a,b are both odd or a,b are both even. The number of pairs a,b with $a < b$ and a,b both even is $50.49/2$; and the number of pairs with $a < b$ and a,b both odd is the same. Hence the total number of triples in arithmetic progression is $50.49 = 2450$.

(b) First count the number of GP's whose common ratio is an integer — that is, GP's k, kn, kn^2 with k, n positive integers such that $n \ge 2$ and $kn^2 \le 100$. Given n, the number of possibilities for k is $[\frac{100}{n^2}]$ (where $[x]$ denotes the largest integer less than or equal to x). Hence the total number of GP's k, kn, kn^2 is

$$\sum_{n=2}^{10} [\frac{100}{n^2}] = 25 + 11 + 6 + 4 + 2 + 2 + 1 + 1 + 1 = 53.$$

Now count the number of GP's $k, km/n, km^2/n^2$, where the common ratio is m/n in lowest terms, with $n > 1$. Here $n^2 | k$, $m > n$ and $km^2 \le 100n^2$. Given n, the number of such GP's is

$$f(n) = \sum_{m>n,(m,n)=1} [\frac{100}{m^2}],$$

and so the total number of GP's with non-integer common ratio is $f(2) + \ldots + f(9)$. Calculate that

$$f(2) = \sum_{m>2, m\ odd} [\frac{100}{m^2}] = 11 + 4 + 2 + 1 = 18,$$

and likewise, $f(3) = 14, f(4) = 7, f(5) = 6, f(6) = 2, f(7) = 3, f(8) = 1, f(9) = 1$.

Hence the total number of GP's in the question is

$$53 + f(2) + \ldots + f(9) = 105.$$

13. (a) The coefficient is $\binom{18}{15} = \frac{18.17.16}{3.2.1} = 816$.

(b) A typical term in the expansion is $\binom{8}{a}(2x^3)^a(\frac{-1}{x^2})^{8-a}$. To make this a term in x^4 we need $3a - 2(8 - a) = 4$, hence $a = 4$. So the coefficient of x^4 is

$$\binom{8}{4}.2^4.(-1)^4 = 1120.$$

(c) A typical term is

$$\binom{10}{a,b,c} y^a (x^{2b}) (\frac{-1}{xy})^c$$

where $a + b + c = 10$. For this to be a constant term, we need $a = c = 2b$, hence $b = 2, a = c = 4$. So the constant term is $\binom{10}{4,2,4} = 3150$.

15. (a) Since p appears in the prime factorization of r, but not of s, it appears in the prime factorization of $\frac{r}{s}$.

(b) For $1 \le k \le p - 1$, $\binom{p}{k} = \frac{p(p-1)\cdots(p-k+1)}{k(k-1)\cdots 1}$ has p dividing the numerator but not the denominator, hence is divisible by p, by part (i).

(c) Let $P(n)$ be the statement that $n^p - n$ is divisible by p. Obviously $P(1)$ is true.

Assume $P(n)$ is true. Then $n^p \equiv n \bmod p$. By the Binomial Theorem,

$$(n+1)^p = n^p + \binom{p}{1}n^{p-1} + \binom{p}{2}n^{p-2} + \cdots + \binom{p}{p-1}n + 1.$$

By (b), all the binomial coefficients $\binom{p}{1}, \binom{p}{2}, \ldots \binom{p}{p-1}$ are divisible by p, so it follows that $(n+1)^p \equiv n^p + 1 \bmod p$. Since $n^p \equiv n \bmod p$, we therefore have $(n+1)^p \equiv n + 1 \bmod p$, which is $P(n+1)$. Hence $P(n) \Rightarrow P(n+1)$, and the result is proved by induction.

17. As you can see from the formula in the question, the 2^{n-1} conjecture is in fact false. (You would have seen this if you had calculated the next value r_6, which is 31, rather than 32.)

The proof of the formula for r_n is a beautiful application of Euler's theorem 9.2 for connected plane graphs. The graph in question is that having vertices consisting of the n points on the circle, together with the points inside the circle where the lines intersect; two vertices are joined by an edge in the graph if they lie on one of the straight lines joining two of the n points on the circle.

The number of regions r_n is equal to the number f of faces of this plane graph plus n (the latter being the n outer faces bounded by an arc of the circle). To calculate f we work out v and e, the numbers of vertices and edges of the graph.

First, for each set of 4 of the n points, the lines joining all pairs of these 4 points intersect in a single point inside the circle, and all vertices of the graph inside the circle arise in this way. Hence $v = n + \binom{n}{4}$.

To calculate e, we count all pairs (x, y), where x is a vertex and y is an edge containing x. If x is one of the $\binom{n}{4}$ interior vertices, then x lies on 4 edges; and if x is one of the n vertices on the circle, then x lies on $n - 1$ edges. Hence the total number of such pairs (x, y) is equal to

$$4\binom{n}{4} + n(n-1).$$

On the other hand, each edge y contains 2 vertices, so the number of pairs is also equal to $2e$. Thus

$$e = 2\binom{n}{4} + \frac{n(n-1)}{2}.$$

Hence, by Theorem 9.2, the number f of faces is

$$f = e - v + 1 = 2\binom{n}{4} + \frac{n(n-1)}{2} - n - \binom{n}{4} + 1.$$

Hence

$$r_n = f + n = \binom{n}{4} + \frac{n(n-1)}{2} + 1 = \binom{n}{4} + \binom{n}{2} + 1.$$

Chapter 17

1. (a) Suppose $A \cup B = A$. Then $x \in B \Rightarrow x \in A$, hence $B \subseteq A$. Conversely, suppose $B \subseteq A$. Then $x \in A \cup B \Leftrightarrow x \in A$, hence $A \cup B = A$.

(b) $x \in (A - C) \cap (B - C) \Leftrightarrow x \in A$ not C and $x \in B$ not $C \Leftrightarrow x \in A$ and B not C $\Leftrightarrow x \in (A \cap B) - C$.

3. Write $U = \bigcup_{n=1}^{\infty} A_n, I = \bigcap_{n=1}^{\infty} A_n$.

(a) Here $U = \{x \in \mathbb{R} | x > 1\}, I = \emptyset$.

(b) $U = \{x \in \mathbb{R} | 0 < x < \sqrt{2} + 1\}, I = \{x \in \mathbb{R} | 1 < x < \sqrt{2}\}$.

(c) $U = \{x \in \mathbb{R} | x < 1\}, I = \{x \in \mathbb{R} | -1 < x < 0\}$.

(d) $U = \{x \in \mathbb{Q} | \sqrt{2} - 1 \leq x \leq \sqrt{2} + 1\}, I = \emptyset$.

5. Let $S = \{1000, 1001, \ldots, 9999\}$. For $k = 0, 8$ or 9, let A_k be the set of integers in S that have no digit equal to k. Then $A_0 \cup A_8 \cup A_9$ is the set of integers in S that are missing either $0, 8$ or 9, and so the number we are asked for in the question is

$$|S| - |A_0 \cup A_8 \cup A_9| = 9000 - |A_0 \cup A_8 \cup A_9|.$$

We therefore need to calculate $|A_0 \cup A_8 \cup A_9|$, which we shall do using the above equality for $|A \cup B \cup C|$. By the Multiplication Principle 16.1, we have $|A_9| = 8 \times 9 \times 9 \times 9$, since there are 8 choices for the first digit (it cannot be 0 or 9) and 9 for each of the others. Similarly,

$$|A_8| = 8 \times 9 \times 9 \times 9 = 5832, \quad |A_0| = 9 \times 9 \times 9 \times 9 = 6561,$$
$$|A_0 \cap A_8| = |A_0 \cap A_9| = 8 \times 8 \times 8 \times 8 = 4096,$$
$$|A_8 \cap A_9| = 7 \times 8 \times 8 \times 8 = 3584,$$
$$|A_0 \cap A_8 \cap A_9| = 7 \times 7 \times 7 \times 7 = 2401.$$

Therefore, by (17.1),

$$|A_0 \cup A_8 \cup A_9| = 5832 + 5832 + 6561 - 4096 - 4096 - 3584 + 2401 = 8850.$$

Hence, the number of integers in S that have at least one of each of the digits $0, 8$ and 9 is equal to $9000 - 8850 = 150$.

7. For $r \geq 2$ let A_r be the set of rth powers k^r such that $2 \leq k^r \leq 10000$. The question asks us to calculate $|A_2 \cup A_3 \cup A_4 \cup A_5|$. Use the Inclusion–Exclusion Principle. First, A_2 consists of the squares $2^2, \ldots 100^2$, so $|A_2| = 99$; similarly $|A_3| = 20$, $|A_4| = 9$, $|A_5| = 5$. Since $A_r \cap A_s = A_d$ where $d = \operatorname{lcm}(r, s)$, we can easily work out the sizes of the intersections, and they are:

$$|A_2 \cap A_3| = |A_6| = 3, \quad |A_2 \cap A_4| = |A_4| = 9,$$
$$|A_2 \cap A_5| = |A_{10}| = 1, \quad |A_3 \cap A_4| = |A_{12}| = 1,$$
$$|A_3 \cap A_5| = |A_{15}| = 0, \quad |A_4 \cap A_5| = |A_{20}| = 0,$$
$$|A_2 \cap A_3 \cap A_4| = |A_{12}| = 1, \quad |A_2 \cap A_3 \cap A_5| = |A_{30}| = 0,$$
$$|A_2 \cap A_4 \cap A_5| = |A_{20}| = 0, \quad |A_3 \cap A_4 \cap A_5| = |A_{60}| = 0,$$
$$|A_2 \cap A_3 \cap A_4 \cap A_5| = |A_{60}| = 0.$$

Hence Inclusion–Exclusion gives $|A_2 \cup A_3 \cup A_4 \cup A_5| = 120$.

9. (a) The number of subsets of $\{1, 2, \ldots, n\}$ of size r is equal to $\binom{n}{r}$. Hence the total number of subsets of $\{1, 2, \ldots, n\}$ is $\sum_{r=0}^{n} \binom{n}{r}$, which by the equality given in the question, is equal to 2^n.

(b) Here is a proof of Proposition 17.3 by induction. Let $P(n)$ be the statement that $\{1, \ldots, n\}$ has 2^n subsets. The set $\{1\}$ has 2 subsets $\emptyset$ and $\{1\}$, so $P(1)$ is true.

Now suppose $P(n)$ is true, and let $S = \{1, 2, \ldots, n+1\}$ and $T = \{1, 2, \ldots, n\}$. By assumption, T has precisely 2^n subsets. For each subset U of T we get 2 subsets of S, namely U and $U \cup \{n+1\}$; and this gives all the subsets of S. Hence S has twice as many subsets as T, so it has precisely 2^{n+1} subsets. This proves that $P(n) \Rightarrow P(n+1)$, and hence proves the result by induction.

11. This is easy, provided we use Proposition 17.3. Let $m, n \geq 2$ be coprime positive integers, with prime factorizations $m = \prod_1^k p_i^{a_i}$, $n = \prod_1^l q_j^{b_j}$. Since m, n are coprime, all the primes p_i, q_j are distinct and the prime factorization of mn is $mn = \prod_{i,j} p_i^{a_i} q_j^{b_j}$. Hence Proposition 17.3 gives

$$
\begin{aligned}
\phi(mn) &= mn \prod_{i,j} \left(1 - \tfrac{1}{p_i}\right)\left(1 - \tfrac{1}{q_j}\right) \\
&= m \prod_i \left(1 - \tfrac{1}{p_i}\right) \cdot n \prod_j \left(1 - \tfrac{1}{q_j}\right) \\
&= \phi(m)\phi(n).
\end{aligned}
$$

13. This is a famous application of Inclusion–Exclusion. Let A_i be the set of arrangements of $1, \ldots, n$ in which the number i is in position i. Then $D(n)$ consists of the arrangements that are not in any of the sets A_i for $i = 1, \ldots, n$, so

$$|D(n)| = n! - |A_1 \cup \cdots \cup A_n|.$$

We use Inclusion–Exclusion to work out $|A_1 \cup \cdots \cup A_n|$. The arrangements in A_i have i in position i and the other $n-1$ numbers in any order, so $|A_i| = (n-1)!$. For $i \neq j$, the arrangements in $A_i \cap A_j$ have i, j in positions i, j and the other $n-2$ numbers in any order, so $|A_i \cap A_j| = (n-2)!$. Similarly, the size of any intersection of r sets $|A_{i_1} \cap \cdots \cap A_{i_r}| = (n-r)!$. Hence by Inclusion–Exclusion,

$$|A_1 \cup \cdots \cup A_n| = n \cdot (n-1)! - \binom{n}{2}(n-2)! + \binom{n}{3}(n-3)! - \cdots + (-1)^{n-1}\binom{n}{n}.$$

Using the fact that $\binom{n}{r} = n!/r!(n-r)!$, the right-hand side of this equation works out as

$$n!\left(1 - \frac{1}{2!} + \frac{1}{3!} - \cdots + (-1)^{n-1}\frac{1}{n!}\right).$$

The conclusion follows from this.

Chapter 18

1. It is easy to see that (i), (iii), (v) and (viii) are equivalence relations.

Less obvious is that (vi) is an equivalence relation. Clearly $a \sim a$ and $a \sim b \Rightarrow b \sim a$; only transitivity is unclear. So suppose $a \sim b$ and $b \sim c$. Then $ab =$

$m^2, bc = n^2$ where m, n are integers. So $ab^2c = m^2n^2$, whence $ac = m^2n^2/b^2$. It follows that all primes in the factorization of ac appear to an even power; hence ac is a square and $a \sim c$, proving transitivity.

The relation in (ii) is not an equivalence relation, as it is not transitive: for example, if $a = 1, b = 0, c = 1$, then $a \sim b$ and $b \sim c$, but $a \nsim c$.

The relation (iv) is also not transitive.

Finally, (vii) is not symmetric, as, for example, $2 \nsim 2$.

3. The relation in Example 18.1(3) is reflexive and symmetric, but not transitive. The relation $a \sim b \Leftrightarrow a \leq b$ on $\mathbb{R}$ is reflexive and transitive, but not symmetric. And if we define a relation $\sim$ on the set $\{1\}$ by $1 \nsim 1$, this is symmetric and transitive, but not reflexive.

5. (a) Let $S = \{1,2\}$. To specify a relation $\sim$ on S, for each ordered pair $(a,b) \in S \times S$, we have 2 choices: either $a \sim b$ or $a \nsim b$. So the number of different relations is equal to $2^{|S \times S|} = 2^4 = 16$.

(b) Let $S = \{1,2,3\}$. To specify a reflexive, symmetric relation $\sim$ on S, we must define $a \sim a$ for all $a \in S$, and $a \sim b \Leftrightarrow b \sim a$. So we have 2 choices ($a \sim b$ or $a \nsim b$) for each (a,b) with $a < b$, and once we have made these choices, the relation $\sim$ is specified. The number of $(a,b) \in S \times S$ with $a < b$ is 3. Hence the number of relations is $2^3 = 8$.

(c) Let $S = \{1,2,\ldots,n\}$. By the argument for (a), the number of relations on S is equal to $2^{|S \times S|} = 2^{n^2}$.

7. Since $m \sim m+5$ and $m+5 \sim m+10$, transitivity implies that $m \sim m+10$, and hence we see that $m \sim m+5t$ for any $t \in \mathbb{Z}$. Similarly $m \sim m+8s$, and hence $m \sim m+8s+5t$ for any $m, s, t \in \mathbb{Z}$. As 5 and 8 are coprime, there exist integers s, t such that $8s + 5t = 1$, and hence $m \sim m+1$ for any $m \in \mathbb{Z}$. It follows that $m \sim n$ for any $m, n \in \mathbb{Z}$.

The partitions of $\{1,2,3\}$ are: $\{1,2,3\}$; $\{1\},\{2,3\}$; $\{2\},\{1,3\}$; $\{3\},\{1,2\}$; $\{1\},\{2\},\{3\}$. So there are exactly 5 equivalence relations on the set $\{1,2,3\}$. Similarly, there are 15 equivalence relations on $\{1,2,3,4\}$ and there are 52 on $\{1,2,3,4,5\}$.

Chapter 19

1. (i) This is not onto, as $f(x) = (x+1)^2 - 1 \geq -1$ for all $x \in \mathbb{R}$. It is also not 1-1, as, for example, $f(0) = f(-2) = 0$.

(ii) This is onto, but not 1-1 (as, for example, $f(0) = f(2) = 0$).

(iii) This is not onto. But it is 1-1, since if $(x + \sqrt{2})^2 = (y + \sqrt{2})^2$ with

$x, y \in \mathbb{Q}$, then it is not possible that $x + \sqrt{2} = -(y + \sqrt{2})$ (as this would imply that $x + y = -2\sqrt{2}$, which is irrational), so $x + \sqrt{2} = y + \sqrt{2}$, hence $x = y$.

(iv) This is not onto (for example, 7 is not in the image of f); but it is 1-1 since by the Fundamental Theorem of Arithmetic 11.1,

$$f(m,n,r) = f(a,b,c) \Rightarrow 2^m 3^n 5^r = 2^a 3^b 5^c \Rightarrow m = a, n = b, r = c.$$

(v) This is not onto; neither is it 1-1, as $f(2,2,1) = f(1,1,2) = 216$.

(vi) There are just 7 equivalence classes, $cl(0), cl(1), \ldots, cl(6)$, and f sends

$$cl(0) \to cl(1) \to cl(2) \to cl(3) \to cl(4) \to cl(5) \to cl(6) \to cl(0).$$

Hence f is both onto and 1-1.

3. If $y = f(x)$, then $y^2 + y + 3 = x^2 - 3x + 5$, so $y^2 + y - (x^2 - 3x + 2) = 0$. Solving this as a quadratic in y gives $y = x - 2$ or $1 - x$. So any function f such that $f(x) \in \{x - 2, 1 - x\}$ for all x will do.

5. (a) Every integer is congruent to 0,1,2,3 or 4 modulo 5. Take the "pigeons" to be the 6 integers, and the "pigeonholes" to be the numbers 0,1,2,3,4. A pigeon goes into the pigeonhole it is congruent to modulo 5. Since there are 6 pigeons and 5 pigeonholes, two of the pigeons must go to the same hole; so two of our six integers must be congruent modulo 5, which means their difference is divisible by 5.

(b) This is just the same argument as for (a). Every integer is congruent to one of the n numbers $0, 1, 2, \ldots, n - 1$ modulo n. Hence, given $n + 1$ integers, the Pigeonhole Principle shows that two of them must be congruent to the same number modulo n. Their difference is then divisible by n.

(c) Since $0, a_1, a_1 + a_2, \ldots, a_1 + \ldots + a_n$ are $n + 1$ integers, by part (b) there must be two of them whose difference is divisible by n. This difference is the sum of a subset of $\{a_1, \ldots, a_n\}$.

(d) The number of 5-element subsets of S is $\binom{10}{5} = 252$. The sum of 5 numbers in S is at most $50 + 49 + 48 + 47 + 46 = 240$. Hence the Pigeonhole Principle shows that two of the 5-element subsets must have the same sum.

(e) The number of subsets of T is $2^9 = 512$, by Proposition 18.3. The sum of the elements of T is at most $50 + 49 + \ldots + 42 = 414$. Hence the Pigeonhole Principle shows that there are two subsets of T having the same sum; let two such subsets be A, B. Then $A - (A \cap B)$ and $B - (A \cap B)$ are two disjoint subsets of T having the same sum.

(f) Every integer is of the form $2^a x$ for some non-negative integer a and odd integer x. For integers in $\{1, \ldots, 200\}$, the possibilities for the odd integer x are $1, 3, 5, \ldots, 199$, so there are just 100 possibilities. Hence, by the Pigeonhole Principle, if we choose 101 integers between 1 and 200, there will be two of them having the same value of x. These two must be of the form $2^a x$ and $2^b x$. If $a > b$, then $2^b x$ divides $2^a x$ (and vice versa). This completes the proof.

7. (a) There are 5^3 functions from S to T, by Proposition 19.3. To count the number of 1-1 functions f, there are 5 choices for $f(1)$, then 4 choices for $f(2)$ and 3 choices for $f(3)$. So the number of 1-1 functions is $5.4.3 = 60$.

(b) This is easy by the argument in (a): for a 1-1 function f, the number of choices for $f(1)$ is n, the number for $f(2)$ is $n-1$, and so on, up to the number of choices for $f(m)$ being $n-m+1$. So the total number of 1-1 functions is $n(n-1)(n-2)\ldots(n-m+1)$.

Chapter 20

1. I will leave you to complete this table.

3. (a) The possible orders of elements of S_4 are 1 (identity element), 2 (cycle-shapes $(2,1^2),(2^2)$), 3 (cycle-shape $(3,1)$) and 4 (cycle-shape (4)). The numbers of elements of these orders are, respectively, 1, 9, 8 and 6.

(b) $(1^6),(2^2,1^2),(3^2),(3,1^3),(4,2),(5,1)$.

(c) The largest order is 30, the order of an element of cycle-shape $(2,3,5)$.

(d) The element of order 30 in the previous part is an odd permutation. The largest order on an *even* permutation in S_{10} is 21, for the cycle-shape $(7,3)$.

(e) $n = 28$ works: S_{28} has an element of cycle-shape $(2,3,5,7,11)$, of order $2 \cdot 3 \cdot 5 \cdot 7 \cdot 11$ which is greater than 28^2.

5. If $g \in S_n$ is an even permutation, then $sgn(g(12)) = sgn(g)\,sgn(12) = -1$ by Proposition 20.5, so $g(12)$ is an odd permutation. So if E and O denote the sets of even and odd permutations in S_n, respectively, we can define a function $\phi : E \to O$ by $\phi(g) = g(12)$ for $g \in E$.

We claim that ϕ is a bijection. It is 1-1, since

$$\phi(g) = \phi(h) \Rightarrow g(12) = h(12) \Rightarrow g(12)(12)^{-1} = h(12)(12)^{-1} \Rightarrow g = h.$$

To see that ϕ is onto, let $x \in O$. Then $x(12)$ is even and $\phi(x(12)) = x(12)(12) = x$. Hence ϕ is a bijection. By Proposition 20.1, this implies that $|E| = |O|$. Since $|E| + |O| = |S_n| = n!$, it follows that $|E| = |O| = \frac{1}{2}n!$.

7. (a) If $|S| = |T|$ (i.e., $m = n$), then an onto function from S to T is automatically a bijection. Write $S = \{s_1,\ldots,s_n\}$. To specify a bijection $f : S \to T$, there are n choices for $f(s_1)$, then $n-1$ choices for $f(s_2)$, and so on, up to 1 choice for $f(s_n)$. Hence the total number of bijections from S to T is $n!$.

(b) Write $S = \{s_1,\ldots,s_{n+1}\}$. If $f : S \to T$ is onto, then there is a unique pair s_i, s_j of elements of S such that $f(s_i) = f(s_j)$. This pair can be chosen in $\binom{n+1}{2}$ ways. Once the pair is chosen, there are n choices for $f(s_i) = f(s_j)$, then $n-1$

choices for the next $f(s_k)$, and so on. So the total number of onto functions is $\binom{n+1}{2}n!$.

(c) Write $S = \{s_1,\ldots,s_{n+2}\}$ and let $f : S \to T$ be onto. Either there is a unique triple s_i, s_j, s_k of elements of S such that $f(s_i) = f(s_j) = f(s_k)$, or there are two disjoint pairs s_i, s_j and s_k, s_l such that $f(s_i) = f(s_j)$, $f(s_k) = f(s_l)$ and $f(s_i) \neq f(s_k)$. The number of choices of triples is $\binom{n+2}{3}$, and the number of choices of two pairs is $\binom{n+2}{n-2,2,2}$. Hence, counting the numbers of choices in the usual way, we see that the total number of onto functions is $\binom{n+2}{3}n! + \binom{n+2}{n-2,2,2}n!$.

Chapter 21

1. (a) Let $B - A = \{b_1,\ldots,b_k\}$ and note that $A \cup B = A \cup (B - A)$. As A is countable, we can list its elements as $a_1, a_2, a_3, \ldots$. Now list the elements of $A \cup B$ as $b_1,\ldots,b_k, a_1, a_2,\ldots$. Hence $A \cup B$ is countable.

(b) Let $C = B - A$ (so again $A \cup B = A \cup C$). If C is finite then the result follows from (a). So assume C is infinite. We can list the elements of C as $c_1, c_2, c_3,\ldots$ and those of A as $a_1, a_2, a_3,\ldots$. Hence list $A \cup C$ as $a_1, c_1, a_2, c_2, a_3, c_3,\ldots$. So $A \cup C = A \cup B$ is countable.

3. We know $\mathbb{Q}$ is countable by Proposition 21.3. We can use the trick in Example 21.3(2) to deduce that $\mathbb{Q} \times \mathbb{Q}$ is also countable: writing $\mathbb{Q} = \{q_1, q_2,\ldots\}$, the function $f : \mathbb{Q} \times \mathbb{Q} \to \mathbb{N}$ defined by $f(q_i, q_j) = 2^i 3^j$ is 1-1; now use 21.4. So we can list the elements of $\mathbb{Q} \times \mathbb{Q}$ as

$$\mathbb{Q} \times \mathbb{Q} = \{(r_1, r_1'), (r_2, r_2'),\ldots,(r_n, r_n'),\ldots\}.$$

Define complex numbers $z_1 = r_1 + ir_1'$, $z_2 = r_2 + ir_2'$, $\ldots, z_n = r_n + ir_n',\ldots$ These are all the complex numbers having rational real and imaginary parts.

Given any real number $\varepsilon > 0$ and any complex number $w = a + ib$, there are rational numbers r, s as close as we like to a, b — in particular, such that $|a - r| < \frac{1}{2}\varepsilon$ and $|b - s| < \frac{1}{2}\varepsilon$. Then

$$|w - (r + is)|^2 = (a - r)^2 + (b - s)^2 < \frac{1}{4}\varepsilon^2 + \frac{1}{4}\varepsilon^2 = \frac{1}{2}\varepsilon^2.$$

Since r, s are rational, $r + is = z_N$ for some N, and $|w - z_N| < \varepsilon$.

5. (a) Let $p_1, p_2, p_3,\ldots$ be the prime numbers, in ascending order (so $p_1 = 2, p_2 = 3, p_3 = 5$, etc.). We define a function $f : S \to \mathbb{N}$ as follows. For a finite subset $A = \{a_1, a_2,\ldots,a_r\}$ of $\mathbb{N}$ with $a_1 < a_2 < \ldots < a_r$, define

$$f(A) = p_1^{a_1} p_2^{a_2} \cdots p_r^{a_r}.$$

We show that f is 1-1. Let B be a finite subset of $\mathbb{N}$, say $B = \{b_1, \ldots b_s\}$ with $b_1 < b_2 < \ldots < b_s$, and suppose that $f(A) = f(B)$. Then

$$f(A) = f(B) \Rightarrow p_1^{a_1} p_2^{a_2} \ldots p_r^{a_r} = p_1^{b_1} p_2^{b_2} \ldots p_s^{b_s}.$$

By the Fundamental Theorem of Arithmetic 11.1, this implies that $r = s$ and $a_i = b_i$ for all i, so that $A = B$. Hence f is 1-1. It follows from Proposition 21.4 that S is countable.

(b) First we show that the set of all subsets (finite or infinite) of $\mathbb{N}$ is uncountable. Each subset S corresponds to an infinite sequence of 0's and 1's, where we put a 1 in position i if $i \in S$, and a 0 if $i \notin S$. This gives a bijection between the set of all subsets of $\mathbb{N}$ and the set of all infinite sequences of 0's and 1's. You showed (yes, I know you did!) the latter set to be uncountable in Exercise 4.

Hence the set of all subsets of $\mathbb{N}$ is uncountable. This is the union of the set of all finite subsets of $\mathbb{N}$ and the set of all infinite subsets of $\mathbb{N}$. The former set is countable by part (a). If the latter were also countable, then the union would be countable by Exercise 1, which is not the case. Hence the set of all infinite subsets of $\mathbb{N}$ is uncountable.

(c) Let F be the set of all functions from $\mathbb{N}$ to $\mathbb{N}$. Suppose F is countable, say $F = \{f_1, f_2, f_3, \ldots\}$. Define a function $f : \mathbb{N} \to \mathbb{N}$ as follows: $f(1) = 1$ if $f_1(1) \neq 1$, and $f(1) = 0$ if $f_1(1) = 1$; $f(2) = 1$ if $f_2(2) \neq 1$, and $f(2) = 0$ if $f_2(2) = 1$; and so on — in general, $f(n) = 1$ if $f_n(n) \neq 1$, and $f(n) = 0$ if $f_n(n) = 1$. Then $f \in F$, but $f \neq f_n$ for all n, since $f(n) \neq f_n(n)$. This is a contradiction. Hence F is uncountable.

Chapter 22

1. (i) LUB $= 7$, GLB $= -2$.

(ii) The given inequality is equivalent to $(x-3)^2 < (x+7)^2$, which works out as $20x > -40$. Hence the GLB is -2 and there is no upper bound.

(iii) The inequality is $x(x^2 - 3) < 0$, hence either $x < 0, x^2 > 3$ or $x > 0, x^2 < 3$. So the LUB is $\sqrt{3}$ and there is no lower bound.

(iv) The smallest solution of $x^2 = a^2 + b^2$ with x, a, b positive integers is $x = 5, a = 3, b = 4$. So the GLB of the set is 5. There is no upper bound, since we can multiply the equation $5^2 = 3^2 + 4^2$ through by any square to get another solution, so the solutions can get as large as we like.

3. We prove this by contradiction. Suppose that S has two different LUBs, say a and b. So $a \neq b$. Since b is an upper bound and a is a LUB, we know that $b \geq a$ (from the definition of LUB). And since a is an upper bound and b is a

LUB, we likewise know that $a \geq b$. But $b \geq a$ and $a \geq b$ together imply that $a = b$, a contradiction. Hence S cannot have two different LUBs.

The same argument works for GLBs.

5. (a) The set $\{1\}$ has LUB $= 1$.

(b) The set $S = \{x : x \in \mathbb{Q} \text{ and } x^2 < 2\}$ has LUB $= \sqrt{2}$. To see this, observe first that $\sqrt{2}$ is certainly an upper bound. If there is a smaller upper bound, say u, then by Exercise 6 of Chapter 2, there is a rational r such that $u < r < \sqrt{2}$. Then $r^2 < 2$, so $r \in S$ and $r > u$, contradicting the fact that u is an upper bound for S. Hence $\sqrt{2}$ is the LUB.

(c) The set $\{\frac{-\sqrt{2}}{n} : n \in \mathbb{N}\}$ consists of irrationals and has LUB $= 0$.

7. Let $S = \{x : x^3 - x - 1 \leq 0\}$. This set has an upper bound (2, for instance), hence it has a LUB. Let c be this LUB. Observe that $c > 0$ as $0 \in S$ and 0 is not an upper bound for S.

We show that $c^3 - c - 1 = 0$, so that c is a root of the cubic equation of the question. We do this by contradiction. Suppose it is false, so either (1) $c^3 - c - 1 > 0$ or (2) $c^3 - c - 1 < 0$.

Case (1): We find a small positive number α such that $(c - \alpha)^3 - (c - \alpha) - 1 > 0$. If we can do this, it will follow that $c - \alpha$ is an upper bound for S, contradicting the fact that c is the *least* upper bound. To find α, note that $(c - \alpha)^3 - (c - \alpha) - 1 = c^3 - c - 1 - 3c^2\alpha + 3c\alpha^2 - \alpha^3 + \alpha$, which is at least $c^3 - c - 1 - 3c^2\alpha$ provided $\alpha > \alpha^3 > 0$, i.e., provided $0 < \alpha < 1$. So if we choose $\alpha > 0$ such that $\alpha < \frac{c^3-c-1}{3c^2}$ and $\alpha < 1$, then $(c - \alpha)^3 - (c - \alpha) - 1 > 0$, giving a contradiction as explained above.

Case (2): We find a small positive number β such that $(c + \beta)^3 - (c + \beta) - 1 < 0$. If we can do this, it will follow that $c + \beta \in S$, contradicting the fact that c is an upper bound for S. To find β, observe that $(c + \beta)^3 - (c + \beta) - 1 = c^3 - c - 1 + 3c^2\beta + 3c\beta^2 + \beta^3 - \beta$ and this is less than $c^3 - c - 1 + 6c^2\beta$, provided $0 < \beta < 1$ and $\beta < c$. So if we choose $\beta > 0$ such that $\beta < \frac{-(c^3-c-1)}{6c^2}$ and also $\beta < 1$ and $\beta < c$, then $(c + \beta)^3 - (c + \beta) - 1 < 0$, giving a contradiction as explained.

We have shown that cases (1) and (2) both give contradictions. Therefore, $c^3 - c - 1 = 0$.

9. Observe that $T_1 \supseteq T_2 \supseteq T_3 \supseteq \ldots$. Assume T_1 has a lower bound, say l. Then by Exercise 2(b), l is a lower bound for all the sets T_n. Hence the Completeness Axiom implies that every T_n has a GLB. Say $b_n = \text{GLB}(T_n)$. Then $b_1 \leq b_2 \leq b_3 \leq \cdots$ by Exercise 2(c).

(a) For this example $b_n = \text{GLB}(T_n) = n$, and $\{b_1, b_2, \ldots\}$ has no LUB.

(b) Here $b_n = \text{GLB}(T_n) = 0$ for all n. The LUB is 0.

(c) Here $b_n = 1$ for all n. The LUB is 1.

Chapter 23

1. (i) $a_n = \frac{1}{1+(5/n)}$ so by quotient rule, $\lim a_n = 1$.

(ii) Limit is 0: let $\varepsilon > 0$ and choose N so that $\frac{1}{\sqrt{N+5}} < \varepsilon$, i.e., $N > \frac{1}{\varepsilon^2} - 5$. Then $n \geq N \Rightarrow |a_n| < \varepsilon$.

(iii) Not convergent: $a_n = \frac{\sqrt{n}}{1+5/n}$, which is greater than $\frac{1}{2}\sqrt{n}$ for $n > 5$, so whatever a and $\varepsilon > 0$ we choose, there does not exist N such that $|a_n - a| < \varepsilon$ for $n \geq N$.

(iv) Limit is 0: note that $|a_n| \leq \frac{1}{\sqrt{n}}$; so for any $\varepsilon > 0$ we choose $N > \frac{1}{\varepsilon^2}$ so that $|a_n| < \varepsilon$ for $n \geq N$.

(v) Limit is $-\frac{1}{5}$: usual argument using quotient rule.

(vi) $a_n = (-1)^{n+1} + \frac{1}{n}$, not convergent.

(vii) Cunningly observe that $a_n(\sqrt{n+1} + \sqrt{n}) = 1$, so $a_n = \frac{1}{\sqrt{n+1}+\sqrt{n}}$, which has limit 0 by the usual kind of argument.

3. As $c = \text{LUB}(S)$, for any $n \in \mathbb{N} \; \exists s_n \in S$ such that $c - \frac{1}{n} < s_n \leq c$. We show that the sequence (s_n) has limit c. Let $\varepsilon > 0$ and choose $N > \frac{1}{\varepsilon}$. Then for $n \geq N$ we have $|s_n - c| < \frac{1}{n} \leq \frac{1}{N} < \varepsilon$. Hence $\lim s_n = c$.

5. This just means $a_n = a$ for all $n \geq N$.

7. (i) We first show that $a_n \geq \sqrt{2}$ for all $n \geq 2$, by induction. It is true for $n = 2$, since $a_2 = \frac{3}{2} > \sqrt{2}$. For the induction step, assume $a_n \geq \sqrt{2}$. Then by the inequality $x^2 + y^2 \geq 2xy$ (which holds since it is equivalent to $x^2 + y^2 - 2xy = (x-y)^2 \geq 0$), we have $a_n^2 + 2 \geq 2\sqrt{2}a_n$. Hence

$$a_{n+1} = \frac{a_n^2 + 2}{2a_n} \geq \frac{2\sqrt{2}a_n}{a_n} = \sqrt{2}.$$

This establishes that $a_n \geq \sqrt{2}$ for all $n \geq 2$.

Now $a_n - a_{n+1} = a_n - \frac{a_n^2+2}{2a_n} = \frac{a_n^2-2}{2a_n}$. Since $a_n^2 \geq 2$ for $n \geq 2$, this shows that $a_n \geq a_{n+1}$ for $n \geq 2$.

(ii) By Question 6, (a_n) is convergent. Let l be the limit. The sequence $(a_{n+1}) = (\frac{a_n^2+2}{2a_n})$ must also converge to l, and by Proposition 23.2, the limit of this sequence is $\frac{l^2+2}{2l}$. Therefore $l = \frac{l^2+2}{2l}$, hence $l = \pm\sqrt{2}$. As a_n is positive for all n, the limit cannot be negative, so $l = \sqrt{2}$.

9. Loser: wrong, e.g., $(-1)^n$. What a loser.

Polly: wrong (meaningless, and frankly silly, to write $|a_n - \infty|$).

Greta: right! Nice one, Greta!

Ally Wooden: wrong, e.g., the sequence $1, -2, 3, -4, 5, -6, \ldots$. Decent effort Ally, if a little wooden.

Gerry: wrong — there is no sequence at all satisfying Gerry's conditions.

Einstein: right! Nice one, Einstein!

Hawking: wrong, e.g., $a_n = -\frac{1}{n}$.

Richard Thomas: wrong — there is no such sequence. Poor effort, Thomas.

Ivor: wrong, e.g., the sequence $0, 1, 0, 2, 0, 3, \ldots$. Awful, Ivor, but frankly not a surprise.

Chapter 24

1. Let f be the constant function $f(x) = a \; \forall x \in \mathbb{R}$. Let $c \in \mathbb{R}$ and let $\varepsilon > 0$. Define $\delta = \varepsilon$. Then

$$|x - c| < \delta \Rightarrow |f(x) - f(c)| = |a - a| = 0 < \varepsilon.$$

So f is continuous at c for all $c \in \mathbb{R}$.

3. This is sort of obvious, but unpleasant. It doesn't change the question if we divide $p(x)$ by a_n, so we can assume that $a_n = 1$, so $p(x) = x^n + a_{n-1}x^{n-1} + \cdots + a_0$. For $x \neq 0$ define

$$q(x) = \frac{p(x)}{x^n} - 1.$$

Then $|q(x)| = |\frac{a_{n-1}}{x} + \frac{a_{n-2}}{x^2} + \cdots + \frac{a_0}{x^n}|$, and so, by the Triangle Inequality 4.12,

$$|q(x)| \leq |\frac{a_{n-1}}{x}| + |\frac{a_{n-2}}{x^2}| + \cdots + |\frac{a_0}{x^n}|.$$

Let $K = \max(|a_{n-1}|, |a_{n-2}|, \ldots, |a_0|)$. Then

$$|q(x)| \leq K(\frac{1}{|x|} + \frac{1}{|x|^2} + \cdots + \frac{1}{|x|^n}).$$

The term in brackets on the RHS is just a geometric series, and for $|x| > 1$ you can easily see that it is less than $\frac{1}{|x|-1}$. Hence

$$|q(x)| < \frac{K}{|x| - 1} \quad \text{for } |x| > 1.$$

Therefore, if we take $|x| > K + 1$, then $|q(x)| < 1$, which means that $p(x) = x^n(q(x) + 1)$ has the same sign as x^n. As n is odd, we can choose a value $x = a$ to make this sign negative, and a value $x = b$ to make it positive.

Told you it was unpleasant.

5. First consider the case where $f(a) \neq 0$. As $\frac{1}{2}|f(a)| > 0$ and f is continuous at a, $\exists \delta_1 > 0$ such that $|x - a| < \delta_1 \Rightarrow |f(x) - f(a)| < \frac{1}{2}|f(a)|$; this implies that $f(x)$ and $f(a)$ have the same sign, hence that $|f(x) - f(a)| = ||f(x)| - |f(a)||$.

Now let $\varepsilon > 0$ and choose $\delta_2 > 0$ such that $|x - a| < \delta_2 \Rightarrow |f(x) - f(a)| < \varepsilon$. Let $\delta = \min(\delta_1, \delta_2)$. Then $|x - a| < \delta \Rightarrow ||f(x)| - |f(a)|| = |f(x) - f(a)| < \varepsilon$. This shows that $|f(x)|$ is continuous at a provided $f(a) \neq 0$. The case where $f(a) = 0$ is much easier and you should have no trouble with it.

7. The key is to define a fiendishly cunning function and apply the Intermediate Value Theorem to it. Here's the fiendish function: let $g(x) = f(x) - f(x - \frac{1}{2})$. Observe that $g(\frac{1}{2}) = f(\frac{1}{2}) - f(0)$ and $g(1) = f(1) - f(\frac{1}{2})$. So as $f(0) = f(1)$, we have $g(\frac{1}{2}) = -g(1)$. As f is continuous on $\mathbb{R}$, so is the function $x \to f(x - \frac{1}{2})$, and hence so is g (by the sum rule). So as $g(\frac{1}{2})$ and $g(1)$ have opposite signs (or are both 0), the Intermediate Value Theorem tells us that there exists c between $\frac{1}{2}$ and 1 such that $g(c) = 0$. This implies that $f(c) = f(c - \frac{1}{2})$. Pretty fiendish!

9. (i) Putting $x = y = 0$ we get $g(0) = g(0)g(0)$, which implies that $g(0) = 0$ or 1.

(ii) If $g(0) = 0$ then $g(x + 0) = g(x)g(0) = 0$ for all x and there is nothing to prove. So assume $g(0) = 1$. Assume g is continuous at 0. Let $a \in \mathbb{R}$. We'll show g is continuous at a. For any $h \in \mathbb{R}$,

$$g(a + h) - g(a) = g(a)g(h) - g(a) = g(a)(g(h) - g(0)).$$

Let $\varepsilon > 0$. As g is continuos at 0, $\exists \delta > 0$ such that $|h| < \delta \Rightarrow |g(a)||g(h) - g(0)| < \varepsilon$. Hence g is continuous at a.

(iii) If $g(a) = 0$, the above equation shows that $g(a + h) = 0$ for all $h \in \mathbb{R}$. Hence g is constant, and so $g(x) = 0$ for all x.

Chapter 25

1. (i) Yes, this is a group: closure holds, as $|z_1| = |z_2| = 1 \Rightarrow |z_1 z_2| = 1$; associativity holds as complex multiplication is associative; identity is 1; and inverse axiom holds as $|z| = 1 \Rightarrow |z^{-1}| = 1$.

(ii) Yes, this is a group: closure holds since $ab + a + b = -1$ implies a or b is -1; check $(a * b) * c = a * (b * c) = abc + ab + ac + bc + a + b + c$, so $*$ is associative; identity is 0; and inverse of $a \in S$ is $-a/(a + 1)$ — this exists, since $a \neq -1$, and is in S since it is not -1.

(iii) Not a group — it is closed, associative, identity is Liebeck, but inverse axiom fails.

(iv) Not a group — similar to (iii).

(v) Not a group — equation is $(z-1)(z^2+1) = 0$, so set is $\{1, i, -i\}$ which is not closed since $i.i = -1$.

(vi) Not a group — no identity.

(vii) Yes, this is a group: closure holds as $\frac{a}{b} + \frac{c}{d} = \frac{ad+bc}{bd}$, and if b, d are odd, so is bd; associativity holds; identity is 0; inverse of $\frac{a}{b}$ is $-\frac{a}{b}$.

(viii) Not a group: identity must be a, but then b has no inverse.

(ix) Yes, this is a group: easiest proof is to check that the multiplication table is the same as that of the group $\{1, -1\}$ under multiplication, with a instead of 1 and b instead of -1.

3. (i) Since $aa^{-1} = a^{-1}a = e$, the inverse of a^{-1} is a.

(ii) Observe that $(ab)(b^{-1}a^{-1}) = a(bb^{-1})a^{-1} = aea^{-1} = e$, and similarly $(b^{-1}a^{-1})(ab) = e$. Hence the inverse of ab is $(b^{-1}a^{-1})$.

5. (a) Note that $(ab)^2 = a^2b^2 \Rightarrow abab = aabb \Rightarrow ba = ab$ (since we can cancel a on the left and b on the right). Hence G is abelian.

(b) This is tricky. We are given that for some integer i,

$$(ab)^i = a^i b^i, \quad (ab)^{i+1} = a^{i+1}b^{i+1}, \quad (ab)^{i+2} = a^{i+2}b^{i+2}, \quad \forall a, b \in G.$$

Using the first two equations, we get

$$a^{i+1}b^{i+1} = (ab)^{i+1} = (ab)^i ab = a^i b^i ab,$$

and cancelling a^i on the left and b on the right gives (1) $ab^i = b^i a$. Similarly, the second two equations give (2) $ab^{i+1} = b^{i+1}a$. Taking inverses in (1) using Q3(ii), we have $b^{-i}a^{-1} = a^{-1}b^{-i}$. Multipliying (2) by these inverses, we get

$$b^{-i}a^{-1}ab^{i+1} = a^{-1}b^{-i}b^{i+1}a.$$

This works out as $b = a^{-1}ba$, so finally $ab = ba$. Hence G is abelian.

(c) You may be annoyed with me for this one: if you are only given this for $i = 0$ and 1, the equations are $(ab)^0 = a^0 b^0$ and $(ab)^1 = a^1 b^1$ which tell you absolutely nothing.

7. (i) To show that D is a group: closure is clear; for associativity check that

$$((x, \varepsilon) * (y, \delta)) * (z, \gamma) = (x, \varepsilon) * ((y, \delta) * (z, \gamma)) = (xy^\varepsilon z^{\varepsilon\delta}, \varepsilon\delta\gamma);$$

the identity is $(1, 1)$; and the inverse of (x, ε) is $(x^{-\varepsilon}, \varepsilon)$. Clearly $|D| = 8$.

(ii) Note $a^2 = (i, 1) * (i, 1) = (-1, 1)$, so $a^4 = (-1, 1) * (-1, 1) = (1, 1) = e$. And $b^2 = (1, -1) * (1, -1) = e$. Also $a^3 = (-i, 1)$. Consider the list of elements

$$e, a, a^2, a^3, b, ab, a^2b, a^3b.$$

If two of these are equal, then cancelling gives $a^r = b^s$ for some $r \in \{0,1,2,3\}$, $s \in \{0,1\}$. But this is not the case. So the 8 listed elements are distinct, so D consists of precisely these 8 elements.

(iii) Check $ba = (1,-1) * (i,1) = (-i,-1) = a^3b$. So $ba \neq ab$ and D is not abelian.

(iv) As in Example 25.3, one can easily work out all products of elements in the above list using the equations $a^4 = b^2 = e$, $ba = a^3b$. Here is the group table:

	e	a	a^2	a^3	b	ab	a^2b	a^3b
e	e	a	a^2	a^3	b	ab	a^2b	a^3b
a	a	a^2	a^3	e	ab	a^2b	a^3b	b
a^2	a^2	a^3	e	a	a^2b	a^3b	b	ab
a^3	a^3	e	a	a^2	a^3b	b	ab	a^2b
b	b	a^3b	a^2b	ab	e	a^3	a^2	a
ab	ab	b	a^3b	a^2b	a	e	a^3	a^2
a^2b	a^2b	ab	b	a^3b	a^2	a	e	a^3
a^3b	a^3b	a^2b	ab	b	a^3	a^2	a	e

(v) For $n = 0,1,2,3,4$, the number of $x \in D$ such that $x^n = e$ is $8,1,6,1,8$ respectively.

Chapter 26

1. (i) Yes, H is the cyclic subgroup $\langle 37 \rangle$.

(ii) Not a subgroup: closure fails — for example, $(12),(23) \in H$ but the product $(12)(23) = (123) \notin H$.

(iii, iv, v, vii) These are all subgroups — just use Proposition 26.1.

(vi) Not a subgroup, since $e \notin H$.

3. (a) Let G be cyclic, say $G = \langle a \rangle$. For any $a^m, a^n \in G$, we have $a^m a^n = a^{m+n} = a^n a^m$. Hence G is abelian.

(b) The groups V and W in Q2 are both abelian. Neither is cyclic, as all the non-identity elements have order 2, hence they generate cyclic subgroups of size 2.

5. (a) Use Proposition 26.1:

(1) $e \in H$ and $e \in K$ (as they are subgroups), hence $e \in H \cap K$.

(2) $x,y \in H \cap K$ implies $xy \in H$ and $xy \in K$ (as they are subgroups), hence $xy \in H \cap K$.

(3) $x \in H \cap K$ implies $x^{-1} \in H$ and $x^{-1} \in K$ (as they are subgroups....yawn), hence $x^{-1} \in H \cap K$.

Hence $H \cap K$ is a subgroup by Proposition 26.1.

(b) Suppose neither $H \subseteq K$ nor $K \subseteq H$. Then we can pick $x \in H - K$ and $y \in K - H$. So $x, y \in H \cup K$. But I claim that $xy \notin H \cup K$: for if $xy = h \in H$, then $y = x^{-1}h \in H$, contradiction; and if $xy = k \in K$, then $x = ky^{-1} \in K$, contradiction. Hence $xy \notin H \cup K$, so $H \cup K$ is not closed, hence not a subgroup.

(c) Consider the group W in Q2. If we write $a = (12)$, $b = (34)$ then $W = \{e, a, b, ab\}$, and all three non-identity elements a, b, ab have order 2. Let $H = \langle a \rangle$, $K = \langle b \rangle$, $L = \langle ab \rangle$. Then H, K and L all have size 2, and $W = H \cup K \cup L$.

7. (a) The generators of $(\mathbb{Z}, +)$ are 1 and -1.

The generators of C_4 are i and $-i$.

The generators of $C_5 = \langle \omega \rangle$ (where $\omega = e^{2\pi i/5}$) are $\omega, \omega^2, \omega^3, \omega^4$, since each of these has order 5.

The generators of $C_6 = \langle v \rangle$ (where $v = e^{2\pi i/6}$) are v and v^5, since all the other powers of v have order 1, 2 or 3.

(b) Let $C_n = \langle \omega \rangle = \{1, \omega, \omega^2, \ldots, \omega^{n-1}\}$, where $\omega = e^{2\pi i/n}$. I claim that the generators of C_n are precisely the elements ω^r, where $1 \leq r \leq n$ and $\mathrm{hcf}(r, n) = 1$. If we prove this claim, it follows that the number of generators is $\phi(n)$, as required.

Consider ω^r, where $1 \leq r \leq n$. Let $d = \mathrm{hcf}(r, n)$. If $d > 1$ then $(\omega^r)^{n/d} = (\omega^n)^{r/d} = 1$, so the order of ω^r is less than n, and hence ω^r is not a generator of C_n (by Proposition 26.3). On the other hand, if $d = 1$ then by Proposition 10.3, there are integers s, t such that $sr + tn = 1$; then $\omega = \omega^{sr+tn} = (\omega^r)^s \in \langle \omega^r \rangle$, so all powers of ω lie in $\langle \omega^r \rangle$ and so ω^r is a generator of C_n. This proves the claim in the previous paragraph.

9. (a) Closure: let $\bar{x}, \bar{y} \in U(\mathbb{Z}_m)$. Then x and y are coprime to m, so by Exercise 5(c) of Chapter 10, xy is also coprime to m. So $\bar{x}\bar{y} \in U(\mathbb{Z}_m)$.

Associativity: this holds, since $(\bar{x}\bar{y})\bar{z}$ and $\bar{x}(\bar{y}\bar{z})$ are both equal to $\bar{r}$, where $xyz \equiv r \bmod m$.

Identity: this is $\bar{1}$.

Inverses: let $\bar{x} \in U(\mathbb{Z}_m)$. By Proposition 13.6 there is an integer y such that $xy \equiv 1 \bmod m$. Replacing y by its remainder on division by m, we have $\bar{x}\bar{y} = \bar{1}$, so $\bar{y}$ is the inverse of $\bar{x}$.

(b) $U(\mathbb{Z}_9)$ has size $\phi(9) = 6$, and is cyclic with generator $\bar{2}$. $U(\mathbb{Z}_{10})$ has size 4 and is cyclic with generator $\bar{3}$. $U(\mathbb{Z}_{12}) = \{\bar{1}, \bar{5}, \bar{7}, \bar{11}\}$ and is not cyclic as all its elements square to the identity.

(c) The statement is true for $r = 3$, since $U(\mathbb{Z}_8)$ is non-cyclic, as observed in Example 26.8.

Now assume it is true for r, so $U = U(\mathbb{Z}_{2^r})$ is non-cyclic. Now $|U| = \phi(2^r) = 2^{r-1}$, so every element of U has order dividing 2^{r-1} by Corollary 26.1. As U is non-cyclic it has no element of order 2^{r-1} by Proposition 26.3, so it follows that all elements have order dividing 2^{r-2}, so that $\bar{x}^{2^{r-2}} = \bar{1}$ for all

$\bar{x} \in U = U(\mathbb{Z}_{2^r})$. This means that 2^r divides $x^{2^{r-2}} - 1$ for all odd integers x. It follows that 2^{r+1} divides $x^{2^{r-1}} - 1 = (x^{2^{r-2}} - 1)(x^{2^{r-2}} + 1)$ for all odd integers x. This implies that in the group $U(\mathbb{Z}_{2^{r+1}})$ we have $\bar{x}^{2^{r-1}} = \bar{1}$ for all elements $\bar{x}$. So $U(\mathbb{Z}_{2^{r+1}})$ has size $\phi(2^{r+1}) = 2^r$, but has no element of order 2^r. Hence it is not cyclic. This completes the proof by induction.

11. Let G be a group containing elements of orders $1, 2, 3, \ldots, 10$. By Corollary 26.1, $|G|$ must be divisible by each of the numbers $1, 2, \ldots, 10$. The lcm of these numbers is 2520, so this divides $|G|$. On the other hand the cyclic group C_{2520} has elements of each of the required orders (since if ω is a generator and $m | 2520$ then $\omega^{2520/m}$ has order m). So the smallest value of $|G|$ is 2520, and C_{2520} is such a group..

13. For $2^{23} - 1$, Proposition 26.10 shows that the possible prime divisors are $\equiv 1 \bmod 23$. The only such prime less than 100 is 47. Check that 47 does in fact divide $2^{23} - 1$.

Let q be a prime divisor of $3^{13} - 1$. The proof of Proposition 26.10 shows that in the group $\mathbb{Z}_q^*$, the element $\bar{3}$ has order either 1 or 13. In the first case $\bar{3} = \bar{1}$, which implies that $q = 2$. In the second case 13 divides $q - 1$ by Corollary 26.1; the only such primes $q < 100$ are 53 and 79. Check that neither of these divides $3^{13} - 1$. Hence 2 is the only prime divisor less than 100 of $3^{13} - 1$.

Now let q be a prime divisor of $79^{11} - 1$. The argument of the previous paragraph shows that either $\overline{79} = \bar{1}$ in $\mathbb{Z}_q^*$, or 11 divides $q - 1$. In the first case q divides 78, so $q = 2, 3$ or 13. In the second case the possible primes $q < 100$ are $23, 67$ and 89. Check that none of these divides $79^{11} - 1$.

15. The anwer is yes: for example, the period of $\frac{1}{49}$ is equal to $\phi(49) = 42$. To see this you just need to check that the smallest positive integer k such that $10^k \equiv 1 \bmod 49$ is $k = 42$.

17. Since $|G|/|H| = r$, the proof of Lagrange's theorem shows that there are exactly r different right cosets of H in G. Let $x \in G$, and consider the $r + 1$ right cosets $H, Hx, Hx^2, \ldots Hx^r$. Two of these must be the same right coset, say $Hx^i = Hx^j$ with $0 \le i < j \le r$. Then $x^j x^{-i} = x^{j-i} \in H$, which gives the required conclusion.

Further Reading

More on the real numbers and analysis:

K. Binmore, *Mathematical Analysis: A Straightforward Approach,* Cambridge University Press, 1982.

R. Haggarty, *Fundamentals of Mathematical Analysis,* Addison-Wesley, 1993.

More on the complex numbers:

M.R. Spiegel, *Complex Variables,* Schaum's Outline Series, McGraw-Hill, 1974.

More on proofs, induction and Euler's formula:

G. Polya, *Induction and Analogy in Mathematics,* Princeton University Press, 1954.

D.J. Velleman, *How to Prove it: A Structured Approach,* Cambridge University Press, 1994.

More on integers, sets and functions:

P. Eccles, *An Introduction to Mathematical Reasoning,* Cambridge University Press, 1997.

More on number theory:

K.H. Rosen, *Elementary Number Theory and Its Applications,* Addison-Wesley, 1993.

More on abstract algebra:

I.N. Herstein, *Topics in Algebra,* Second Edition, John Wiley and Sons, 1975.

History:

D.M. Burton, *The History of Mathematics: An Introduction,* McGraw-Hill, 1997.

S. Singh, *The Code Book: The Secret History of Codes and Code-Breaking*, Fourth Estate Ltd., 2002.

General:

K. Ball, *Strange Curves, Counting Rabbits and other Mathematical Explorations*, Princeton University Press, 2003.

R. Courant and H. Robbins, *What is Mathematics?,* Oxford University Press, 1979.

Index of Symbols

$A \times B$	Cartesian product of A and B, 149
$\phi(n)$	Euler ϕ-function, 152
$f : S \to T$	function from S to T, 163
$f(S)$	image of function f, 163
ι_S	identity function on S, 164
f^{-1}	inverse function, 166
$g \circ f$	composition of functions, 167
S_n	set of all permutations of $\{1,\ldots,n\}$, 173
ι	identity permutation, 174
Π	product notation, 182
$sgn(g)$	signature of permutation g, 182
$P(S)$	set of all subsets of S, 195
LUB	least upper bound, 200
GLB	greatest lower bound, 200
$\lim a_n$	limit of sequence (a_n), 208
$a_n \to a$	sequence (a_n) has limit a, 208
S_n	symmetric group of degree n, 228
e	identity element of a group, 229
x^{-1}	inverse of element x in a group, 229
C_n	cyclic group of size n, 236
A_n	alternating group of degree n, 237
$o(a)$	order of element a in a group, 239
$\mathbb{Z}_p^*$	group $\mathbb{Z}_p - \{\bar{0}\}$, 243
$U(\mathbb{Z}_m)$	group contained in $\mathbb{Z}_m$, 244

Index